Understanding Diabetes

AF690406

LICENSE, DISCLAIMER OF LIABILITY, AND LIMITED WARRANTY

By purchasing or using this book and companion files (the "Work"), you agree that this license grants permission to use the contents contained herein, including the disc, but does not give you the right of ownership to any of the textual content in the book / disc or ownership to any of the information or products contained in it. *This license does not permit uploading of the Work onto the Internet or on a network (of any kind) without the written consent of the Publisher.* Duplication or dissemination of any text, code, simulations, images, etc. contained herein is limited to and subject to licensing terms for the respective products, and permission must be obtained from the Publisher or the owner of the content, etc., in order to reproduce or network any portion of the textual material (in any media) that is contained in the Work.

Mercury Learning and Information ("MLI" or "the Publisher") and anyone involved in the creation, writing, or production of the companion disc, accompanying algorithms, code, or computer programs ("the software"), and any accompanying Web site or software of the Work, cannot and do not warrant the performance or results that might be obtained by using the contents of the Work. The author, developers, and the Publisher have used their best efforts to ensure the accuracy and functionality of the textual material and/or programs contained in this package; we, however, make no warranty of any kind, express or implied, regarding the performance of these contents or programs. The Work is sold "as is" without warranty (except for defective materials used in manufacturing the book or due to faulty workmanship).

The author, developers, and the publisher of any accompanying content, and anyone involved in the composition, production, and manufacturing of this work will not be liable for damages of any kind arising out of the use of (or the inability to use) the algorithms, source code, computer programs, or textual material contained in this publication. This includes, but is not limited to, loss of revenue or profit, or other incidental, physical, or consequential damages arising out of the use of this Work.

The sole remedy in the event of a claim of any kind is expressly limited to replacement of the book and/or disc, and only at the discretion of the Publisher. The use of "implied warranty" and certain "exclusions" varies from state to state and might not apply to the purchaser of this product.

UNDERSTANDING DIABETES
A Comprehensive Q&A Guide

Austin Mardon, Ph.D.

MERCURY LEARNING AND INFORMATION
Boston, Massachusetts

Copyright ©2026 by Mercury Learning and Information.
An Imprint of DeGruyter Brill Inc. All rights reserved.

This publication, portions of it, or any accompanying software may not be reproduced in any way, stored in a retrieval system of any type, or transmitted by any means, media, electronic display, or mechanical display, including, but not limited to, photocopy, recording, Internet postings, or scanning, without prior permission in writing from the publisher.

Mercury Learning and Information
121 High Street, 3rd Floor
Boston, MA 02110
info@merclearning.com

A. Mardon. *Understanding Diabetes: A Comprehensive Q&A Guide*
ISBN: 978-1-5015-2754-8

The publisher recognizes and respects all marks used by companies, manufacturers, and developers as a means to distinguish their products. All brand names and product names mentioned in this book are trademarks or service marks of their respective companies. Any omission or misuse (of any kind) of service marks or trademarks, etc. is not an attempt to infringe on the property of others.

Library of Congress Control Number: 2025949246

242526321 This book is printed on acid-free paper in the United States of America.

Our titles are available for adoption, license, or bulk purchase by institutions, corporations, etc.

All of our titles are available in digital format at various digital vendors.

Contents

Preface

Diabetes is a complex and often misunderstood condition that affects millions of people worldwide. Its impact extends beyond individual health to encompass families, communities, and healthcare systems globally. This book serves as a beacon of clarity amidst the complexities of diabetes, aiming to provide clear, concise, and accessible answers to the myriad questions that surround this condition.

Whether you are newly diagnosed, caring for someone with diabetes, or simply seeking to deepen your understanding, this guide will serve as a valuable companion on your journey. Each chapter is dedicated to addressing fifty of the most pertinent questions about diabetes, meticulously crafted to offer comprehensive insights into its various facets. From understanding the underlying causes and types of diabetes to exploring effective management strategies and navigating the emotional and practical challenges it presents, this book strives to empower you with the knowledge and resources needed to navigate the complexities of diabetes with confidence.

The goal of this volume is not only to inform but also to empower. By equipping you with accurate information and practical guidance, we aim to foster a sense of empowerment in managing and living well with diabetes. We recognize that diabetes management is a dynamic journey that requires ongoing learning, adaptation, and support. Through this book, we hope to bridge the gap between medical knowledge and personal empowerment, enabling you to make informed decisions about your health and well-being.

As you embark on this exploration of diabetes, we encourage you to approach each chapter with curiosity and an open mind. May this guide serve as a trusted companion, providing clarity, reassurance, and practical tools to enhance your journey with diabetes. Together, let us embark on a journey toward understanding, managing, and thriving with diabetes.

Introduction to Diabetes

CHAPTER 1

Introduction

Diabetes stands as a formidable global health challenge, impacting millions of lives with its intricate web of causes, types, and management complexities. Defined by elevated blood sugar levels, diabetes encompasses diverse forms—Type 1, Type 2, gestational, and diabetes insipidus—each demanding tailored approaches to diagnosis, treatment, and long-term care. Understanding the distinct mechanisms behind each type unveils the varied interplay of genetic predispositions, autoimmune responses, insulin resistance, and hormonal fluctuations.

Type 1 diabetes arises from an autoimmune attack on insulin-producing cells, necessitating lifelong insulin therapy and vigilant management. In contrast, Type 2 diabetes typically develops from a combination of insulin resistance and inadequate insulin production, often influenced by lifestyle factors such as sedentary habits and poor dietary choices. Gestational diabetes manifests during pregnancy, underscoring the temporary yet critical need for controlled blood sugar levels to safeguard maternal and fetal health. Meanwhile, diabetes insipidus disrupts fluid balance due to impairments in either the brain's production of antidiuretic hormone or the kidneys' response to it, presenting distinctive challenges in diagnosis and management.

Effective diabetes management integrates advanced diagnostic tools like continuous glucose monitors with personalized treatment plans encompassing medication regimens, dietary adjustments, and regular physical activity. The pursuit of optimal health outcomes involves not only meticulous blood sugar monitoring and medication adherence but also proactive efforts in preventing complications through lifestyle modifications and comprehensive healthcare support.

This introduction sets the stage for exploring the intricate landscape of diabetes—its origins, complexities, management strategies, and the crucial role of patient empowerment and healthcare collaboration in mitigating its impact on individuals and society at large.

▲ Figure 1.1
Diabetes infographic.

1. What is diabetes?

Diabetes is a chronic condition characterized by high levels of glucose (sugar) in the blood due to the body's inability to produce or properly use insulin. It can lead to serious health complications if not managed properly.

2. What are the main types of diabetes?

The main types of diabetes are:

1. **Type 1 Diabetes Mellitus:** An autoimmune condition where the body attacks insulin-producing beta cells in the pancreas.
2. **Type 2 Diabetes Mellitus:** A metabolic disorder where the body becomes resistant to insulin or doesn't produce enough insulin.
3. **Diabetes Insipidus:** an uncommon problem that affects the fluid volume in the body to become out of balance and cause a large amount of urine production.
4. **Gestational Diabetes Mellitus:** A form of diabetes that develops during pregnancy and usually resolves after childbirth.

3. What causes Type 1 versus Type 2 diabetes?

Type 1 diabetes is caused by an autoimmune response when the immune system mistakenly attacks and destroys insulin-producing beta cells in the pancreas. The exact cause is unknown, but it is believed to involve genetic and environmental factors.

Type 2 diabetes is primarily caused by insulin resistance, when the body's cells do not respond properly to insulin. This is often coupled with an eventual decline in insulin production by the pancreas. Contributing factors include genetics, obesity, physical inactivity, and poor diet.

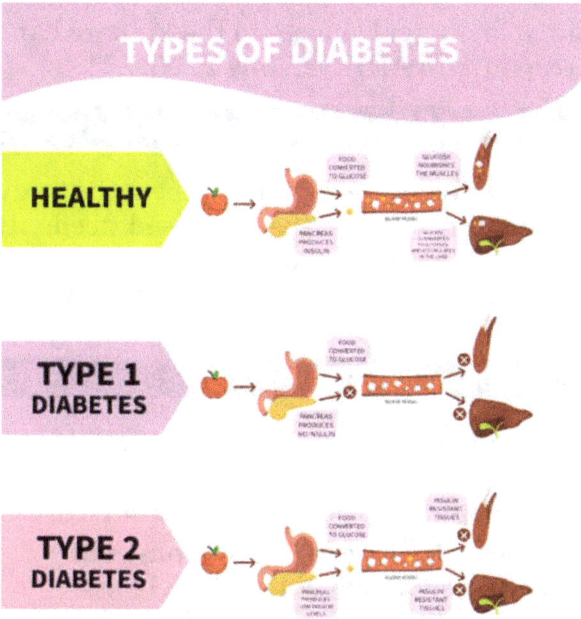

▲ Figure 1.2
Type 1 versus Type 2 diabetes.

4. What causes diabetes insipidus?

There are two causes of diabetes insipidus: central and nephrogenic. Central refers to it being caused by the brain, and nephrogenic refers to it being caused by the kidneys.

1. **Central Diabetes Insipidus (CDI):**
   - **Damage to the hypothalamus or pituitary gland:** Conditions such as tumors, surgery, infections, or head injuries can damage these areas, disrupting the production, release, or function of ADH.
   - **Idiopathic:** In some cases, the cause of CDI remains unknown.

2. **Nephrogenic Diabetes Insipidus (NDI):**
   - **Genetic mutations:** Inherited genetic mutations can affect the kidneys' ability to respond to ADH.
   - **Certain medications:** Drugs such as lithium, used in psychiatric disorders, can impair kidney function related to ADH.
   - **Electrolyte imbalances:** Conditions like high calcium levels (hypercalcemia) can interfere with ADH action.
   - **Chronic kidney disease:** Damage to the kidneys can disrupt their response to ADH, leading to NDI.

5. How is gestational diabetes different from other types?

Gestational diabetes occurs during pregnancy when the body cannot produce enough insulin to meet the increased needs. It usually disappears after childbirth, but women who have had gestational diabetes are at higher risk of developing Type 2 diabetes later in life.

6. Who is at risk for developing diabetes?

Risk factors for developing diabetes include:

- **Type 1 diabetes:** Family history, genetic predisposition, and possibly exposure to certain viruses.
- **Type 2 diabetes:** Family history, obesity, physical inactivity, poor diet, age (45 and older), high blood pressure, and certain ethnic backgrounds (e.g., African American, Hispanic, Native American, Asian American, Pacific Islander).
- **Gestational diabetes:** Obesity, family history of diabetes, previous gestational diabetes, and certain ethnic backgrounds.

7. What are the common symptoms of diabetes?

Common symptoms include:

- Increased thirst and dry mouth
- Frequent urination
- Extreme hunger
- Unexplained weight loss
- Fatigue and weakness
- Blurred vision

- Slow healing of cuts and bruises
- Tingling or numbness in hands and feet

8. How is diabetes diagnosed?

Diabetes is diagnosed through several blood tests, including:

- **Fasting Plasma Glucose Test (FPG):** Measures blood glucose after an overnight fast.
- **A1C Test:** Measures average blood glucose over the past 2– 3 months.
- **Oral Glucose Tolerance Test (OGTT):** Measures blood glucose before and after drinking a sugary drink.
- Measures blood glucose at any time without fasting.

9. Can diabetes be cured?

There is currently no cure for diabetes. However, it can be effectively managed with lifestyle changes, medication, and regular monitoring. Management strategies include healthy eating, regular physical activity, maintaining a healthy weight, and taking prescribed medications.

10. How does insulin work in the body?

Insulin is a hormone produced by the pancreas that allows cells to take in glucose from the blood to be used for energy. It helps regulate blood sugar levels by signaling cells to absorb glucose, which lowers blood glucose levels and provides the body with the energy it needs to function.

11. What happens if diabetes is left untreated?

If diabetes is left untreated, it can lead to a range of serious complications. High blood sugar levels over time can cause damage to blood vessels and nerves, leading to issues such as cardiovascular disease, kidney failure, neuropathy (nerve damage), retinopathy (eye damage), and increased risk of infections. Chronic high blood sugar can also lead to diabetic ketoacidosis, a potentially life-threatening condition.

12. Can diabetes be prevented?

Type 1 diabetes cannot be prevented as it is an autoimmune condition. However, Type 2 diabetes can often be prevented or delayed through healthy lifestyle choices. Maintaining a balanced diet, engaging in regular physical activity, achieving and maintaining a healthy weight, and avoiding smoking are key strategies to reduce the risk of developing Type 2 diabetes.

13. How does diet affect diabetes?

Diet plays a crucial role in managing diabetes by helping to regulate blood sugar levels. Consuming balanced meals with appropriate portions of carbohydrates, proteins, and fats is essential. Carbohydrates have the most direct impact on blood sugar levels, so monitoring and managing carb intake is important. A diet rich in whole grains, vegetables, lean proteins, and healthy fats can help maintain stable blood sugar levels and overall health.

14. What are the different types of insulin?

There are several types of insulin, categorized by how quickly they start to work and how long they last in the body:

- **Rapid-acting insulin:** Starts working within 15 minutes and lasts for 2–4 hours.
- **Short-acting insulin:** Starts working within 30 minutes and lasts for 3–6 hours.
- **Intermediate-acting insulin:** Starts working within 2–4 hours and lasts for 12– 18 hours.
- **Long-acting insulin:** Starts working within a few hours and lasts for 24 hours or more.
- **Ultra-long-acting insulin:** Lasts for up to 36 hours or more, providing a steady level of insulin.

15. How is insulin administered?

Insulin is typically administered through injections or an insulin pump. Injections can be given using a syringe, insulin pen, or insulin jet injector. An insulin pump delivers a continuous infusion of insulin through a small tube inserted under the skin, allowing for precise control of blood sugar levels. The method of administration depends on the individual's needs, preferences, and medical advice.

16. What are oral diabetes medications?

Oral diabetes medications are primarily used to manage Type 2 diabetes. They include:

- **Metformin:** Reduces glucose production in the liver and improves insulin sensitivity.
- **Sulfonylureas:** Stimulate the pancreas to release more insulin.
- **DPP-4 inhibitors:** Help increase insulin production and decrease glucose production.
- **SGLT2 inhibitors:** Help the kidneys remove glucose from the bloodstream.
- **Thiazolidinediones:** Improve insulin sensitivity. These medications work in different ways and may be used alone or in combination to achieve optimal blood sugar control.

17. Can people with diabetes lead a normal life?

Yes, with proper management, people with diabetes can lead healthy and active lives. Effective diabetes management involves regular monitoring of blood sugar levels, adhering to a balanced diet, engaging in regular physical activity, taking prescribed medications, and attending regular medical checkups. Education and support from healthcare professionals can help individuals with diabetes navigate challenges and maintain a high quality of life.

18. How does sleep affect diabetes?

Sleep plays a crucial role in overall health and blood sugar control. Poor sleep or insufficient sleep can lead to insulin resistance, increased appetite, and higher blood sugar levels, which can exacerbate diabetes symptoms. Consistently getting 7–9 hours of quality sleep per night is important for managing diabetes effectively. Good sleep hygiene practices, such as maintaining a regular sleep schedule, creating a restful sleep environment, and avoiding caffeine or electronics before bedtime, can significantly improve sleep quality and, consequently, blood sugar management.

19. What is diabetic ketoacidosis (DKA)?

Diabetic ketoacidosis (DKA) is a serious and potentially life-threatening complication of diabetes, more commonly seen in Type 1 diabetes but can also occur in Type 2. It occurs when the body lacks sufficient insulin to allow glucose into cells for energy, causing the body to start breaking down fats too quickly. This process produces ketones, which are acids that accumulate in the blood, leading to acidosis. Symptoms of DKA include high blood sugar levels, high ketone levels in urine, frequent urination, excessive thirst, nausea, vomiting, abdominal pain, shortness of breath, and confusion. Immediate medical attention is required to treat DKA and prevent severe complications or death.

20. What is hypoglycemia?

Hypoglycemia is a condition characterized by abnormally low blood sugar levels, typically below 70 mg/dL. It can occur in individuals with diabetes who use insulin or other glucose-lowering medications. Symptoms of hypoglycemia include shakiness, sweating, rapid heartbeat, confusion, irritability, dizziness, and fainting. Severe hypoglycemia can lead to seizures, unconsciousness, and even death if not treated promptly. Treatment involves consuming fast-acting carbohydrates, such as glucose tablets, fruit juice, or regular soda, followed by a more substantial snack or meal to stabilize blood sugar levels.

21. What is hyperglycemia?

Hyperglycemia is a condition where blood sugar levels are too high, typically above 180 mg/dL after meals or above 130 mg/dL when fasting. It can occur when the body has insufficient insulin or cannot use insulin effectively. Symptoms include frequent urination, increased thirst, fatigue, blurred vision, and headaches. If left untreated, hyperglycemia can lead to serious complications such as cardiovascular disease, nerve damage, kidney damage, and vision problems. Managing hyperglycemia involves monitoring blood sugar levels, adhering to prescribed medications, maintaining a healthy diet, and regular physical activity.

22. Can diabetes affect mental health?

Yes, diabetes can significantly impact mental health. The constant management of the condition, concerns about complications, and the effects of blood sugar fluctuations can increase the risk of depression, anxiety, and diabetes-related distress. Chronic stress and mental health disorders can, in turn, make it more challenging to manage diabetes effectively. Seeking support from healthcare professionals, mental health counselors, support groups, and utilizing stress management techniques are essential strategies for maintaining both physical and mental health in individuals with diabetes.

23. What is the glycemic index?

The glycemic index (GI) is a measure that ranks foods containing carbohydrates based on how quickly they raise blood sugar levels after consumption. Foods are scored on a scale from 0 to 100, with higher values indicating faster increases in blood sugar levels. Low-GI foods (55 or less) cause slower, more gradual rises in blood sugar, whole high-GI foods (70 or more) cause rapid spikes. Managing diabetes involves choosing low to moderate-GI foods to help maintain stable blood sugar levels. Examples of low-GI foods include whole grains, legumes, nonstarchy vegetables, and most fruits. Some high-GI foods include white bread and bagels, white rice, sugary breakfast cereals (such as corn flakes), and even watermelon

24. How can fiber help in managing diabetes?

Fiber, particularly soluble fiber, can help control blood sugar levels by slowing the absorption of sugar into the bloodstream. It also aids in digestion, helps maintain a healthy weight, and lowers cholesterol levels. High-fiber foods include whole grains, fruits, vegetables, legumes, nuts, and seeds. For individuals with diabetes, incorporating fiber into meals can lead to better blood sugar control, reduced insulin resistance, and improved overall health. The recommended daily intake of fiber is 25 grams for women and 38 grams for men.

25. What are the benefits of regular health checkups for diabetics?

Regular health checkups are crucial for individuals with diabetes as they help in the early detection and management of complications. These checkups typically include monitoring blood sugar levels, blood pressure, cholesterol levels, kidney function, and eye health. Regular visits to healthcare providers enable timely adjustments to treatment plans, provide opportunities for receiving updated education on diabetes management, and offer support for lifestyle changes. Early detection of complications can lead to more effective interventions and better long-term outcomes.

26. How does alcohol consumption affect diabetes?

Alcohol can affect blood sugar levels and interact with diabetes medications. Moderate alcohol consumption can cause blood sugar levels to fluctuate, leading to hypoglycemia, especially when consumed on an empty stomach. Alcohol can also mask symptoms of low blood sugar, making it harder to recognize and treat. For individuals with diabetes, it is recommended to consume alcohol in moderation, limit intake to no more than one drink per day for women and two drinks per day for men, and monitor blood sugar levels closely when drinking. It's important to discuss alcohol consumption with a healthcare provider to understand individual risks and guidelines.

27. What is diabetic neuropathy?

Diabetic neuropathy is a type of nerve damage that can occur in individuals with diabetes, resulting from high blood sugar levels over time. It most commonly affects the legs and feet, leading to symptoms such as pain, tingling, numbness, and loss of sensation. There are different types of diabetic neuropathy, including peripheral neuropathy, autonomic neuropathy, proximal neuropathy, and focal neuropathy, each affecting different parts of the body. Managing blood sugar levels, maintaining a healthy lifestyle, and regular foot care are essential strategies to prevent and manage diabetic neuropathy. In severe cases, medication may be needed to manage pain and other symptoms.

28. What are the signs of diabetic retinopathy?

Diabetic retinopathy is a complication of diabetes that affects the eyes. Early signs may be mild and go unnoticed, but as the condition progresses, symptoms can include:

- Blurred or fluctuating vision
- Dark spots or "floaters" in the vision

- Difficulty seeing at night
- Impaired color vision
- Vision loss in severe cases

Regular eye exams are crucial for early detection and management of diabetic retinopathy to prevent vision loss.

29. How can diabetics care for their feet?

Proper foot care is essential for people with diabetes to prevent complications such as infections and ulcers. Key steps include:

- Inspecting feet daily for cuts, blisters, redness, or swelling
- Washing feet daily with lukewarm water and drying thoroughly, especially between the toes
- Moisturizing feet to prevent dry skin, but avoiding the area between the toes
- Wearing comfortable, well-fitting shoes and clean, dry socks
- Avoiding walking barefoot, even at home
- Trimming toenails straight across and filing the edges
- Seeking medical attention promptly for any sores, infections, or foot issues

▲ **Figure 1.3**
Healthy foot routine.

30. What is the role of a diabetes educator?

A diabetes educator is a healthcare professional who specializes in helping individuals manage their diabetes effectively. Their role includes:

- Providing education on diabetes management, including diet, exercise, and medication
- Offering support and counseling to address emotional and psychological aspects of diabetes
- Helping individuals set and achieve personalized health goals
- Teaching self-monitoring of blood glucose levels and interpreting the results
- Assisting with problem-solving and decision-making to improve diabetes control
- Coordinating care with other healthcare providers

31. Can children develop diabetes?

Yes, children can develop both Type 1 and Type 2 diabetes. Type 1 diabetes is more common in children and involves an autoimmune attack on insulin-producing cells. Type 2 diabetes, traditionally seen in adults, is increasingly being diagnosed in children due to rising obesity rates and sedentary lifestyles. Early diagnosis and management are crucial to prevent complications and ensure healthy growth and development.

32. How does pregnancy affect diabetes?

Pregnancy can complicate diabetes management due to hormonal changes that affect insulin sensitivity. Women with pre-existing diabetes (Type 1 or Type 2) need careful monitoring and adjustment of their treatment plans. Additionally, some women may develop gestational diabetes during pregnancy, which usually resolves after childbirth but increases the risk of developing Type 2 diabetes later. Proper prenatal care and management are vital to ensure a healthy pregnancy and reduce risks for both mother and baby.

33. What are the long-term complications of diabetes?

Long-term complications of diabetes arise from prolonged high blood sugar levels and include:

- **Cardiovascular disease:** Increased risk of heart attack, stroke, and peripheral artery disease
- **Kidney disease (nephropathy):** Potentially leading to kidney failure and the need for dialysis or transplantation
- **Nerve damage (neuropathy):** Causing pain, tingling, and loss of sensation, particularly in the extremities
- **Eye damage (retinopathy):** Leading to vision loss and blindness
- **Foot problems:** Including ulcers, infections, and, in severe cases, amputation
- **Skin conditions:** Increased susceptibility to infections and other skin problems
- **Hearing impairment:** Increased risk of hearing loss

34. How does smoking affect diabetes?

Smoking significantly increases the risk of complications in individuals with diabetes, including:

- **Cardiovascular disease:** Higher likelihood of heart attack, stroke, and peripheral artery disease
- **Neuropathy:** Worsened nerve damage and pain
- **Nephropathy:** Accelerated progression of kidney disease
- **Retinopathy:** Increased risk of severe eye complications
- **Poor blood circulation:** Leading to slower healing of wounds and higher risk of infections
- **Respiratory issues:** Greater susceptibility to respiratory infections and reduced lung function

Quitting smoking is one of the most important steps people with diabetes can take to improve their overall health and reduce complications.

35. Can pets get diabetes?

Yes, pets, particularly dogs and cats, can develop diabetes. Symptoms in pets may include increased thirst, frequent urination, weight loss despite a good appetite, lethargy, and in some cases, vision problems. Diabetes in pets is managed similarly to humans, with insulin therapy, dietary changes, and regular monitoring of blood glucose levels. Early diagnosis and proper management are essential to ensure a good quality of life for diabetic pets.

36. What is the role of genetics in diabetes?

Genetics play a significant role in the risk of developing both Type 1 and Type 2 diabetes. For Type 1 diabetes, genetic factors combined with environmental triggers (such as viral infections) can lead to an autoimmune response that destroys insulin-producing cells. In Type 2 diabetes, family history and genetic predisposition increase the risk, particularly when combined with lifestyle factors such as obesity and physical inactivity. Understanding genetic risks can help in early detection and prevention strategies.

37. What is the importance of hydration in diabetes management?

Staying hydrated is crucial for maintaining healthy blood sugar levels and overall health. Proper hydration helps the kidneys flush out excess glucose through urine and prevents dehydration, which can cause blood sugar levels to rise. Drinking plenty of water and avoiding sugary drinks is recommended. Individuals with diabetes should monitor their fluid intake, especially during hot weather or physical activity, to maintain optimal hydration.

38. Can diabetes affect dental health?

Yes, diabetes can increase the risk of dental problems, including:

- **Gum disease (periodontitis):** Higher blood sugar levels can lead to more plaque buildup, causing gum inflammation and infection
- **Dry mouth (xerostomia):** Leading to an increased risk of cavities and oral infections
- **Thrush:** A fungal infection more common in people with diabetes

- **Slow healing:** Wounds in the mouth may take longer to heal. Maintaining good oral hygiene, regular dental checkups, and controlling blood sugar levels are important for preventing dental complications.

39. What is the relationship between diabetes and cholesterol?

People with diabetes often have issues with cholesterol levels, which can increase the risk of cardiovascular disease. This includes:

- **High LDL (bad) cholesterol:** Leading to plaque buildup in arteries
- **Low HDL (good) cholesterol:** Reducing the protective effects against heart disease
- **High triglycerides:** Further increasing the risk of cardiovascular complications. Managing cholesterol levels through diet, exercise, medication, and blood sugar control is essential to reduce cardiovascular risk.

40. How does menopause affect diabetes?

Menopause can affect blood sugar control and increase the risk of diabetes complications. Hormonal changes during menopause can cause fluctuations in blood sugar levels and affect insulin sensitivity. Symptoms such as hot flashes, sleep disturbances, and mood changes can also impact diabetes management. Postmenopausal women may face increased risks of cardiovascular disease and osteoporosis. Regular monitoring, adjusting diabetes treatment plans, and maintaining a healthy lifestyle are important during this time.

41. Can alternative therapies help in managing diabetes?

Some alternative therapies may help in managing diabetes, but they should be used in conjunction with conventional treatment and under medical supervision. These therapies include:

- **Herbal supplements:** Cinnamon, fenugreek, and bitter melon may help with blood sugar control
- **Acupuncture:** To relieve pain and improve overall well-being
- **Yoga and Tai Chi:** For stress reduction and improved physical health

- **Mindfulness and meditation:** To reduce stress and enhance emotional health, It is essential to discuss any alternative therapies with a healthcare provider to ensure they are safe and effective.

42. What is the impact of climate on diabetes management?

Extreme temperatures can affect blood sugar levels and diabetes medications. In hot weather, dehydration and heat exhaustion can cause blood sugar levels to rise or fall unpredictably. Cold weather can lead to reduced physical activity and increased appetite, impacting blood sugar control. Diabetes medications and supplies, such as insulin and glucose monitors, may be affected by extreme temperatures. It is important to take precautions, stay hydrated, and monitor blood sugar levels more frequently during extreme weather conditions.

43. How can support groups help people with diabetes?

Support groups provide emotional support, information, and encouragement for people managing diabetes. Benefits include:

- **Sharing experiences and tips:** Learning from others who face similar challenges
- **Emotional support:** Reducing feelings of isolation and providing a sense of community
- **Education:** Access to resources and information about diabetes management
- **Motivation:** Encouraging adherence to treatment plans and healthy lifestyle choices
- **Problem-solving:** Finding practical solutions to common issues related to diabetes. Support groups can be found in local communities, hospitals, or online.

44. What is insulin resistance?

Insulin resistance is a condition where the body's cells do not respond properly to insulin, leading to higher blood sugar levels. It is a key feature of Type 2 diabetes and can also occur in people with prediabetes. Factors contributing to insulin resistance include obesity, physical inactivity, genetics, and certain medical conditions. Managing insulin resistance involves lifestyle changes such as weight loss, increased physical activity, and a healthy diet, along with medications if necessary.

45. How can mindfulness and meditation help in managing diabetes?

Mindfulness and meditation can reduce stress and improve overall well-being, aiding in better diabetes management. Benefits include:

- **Reduced stress:** Lowering cortisol levels, which can positively impact blood sugar control
- **Improved mental health:** Decreasing symptoms of depression and anxiety
- **Enhanced focus:** Helping individuals stay on track with their diabetes management plans
- **Better lifestyle choices:** Encouraging mindful eating and increased physical activity Incorporating mindfulness and meditation practices into daily routines can support effective diabetes management.

46. What resources are available for people with diabetes?

Numerous resources are available to support individuals with diabetes, including:

- **Healthcare providers:** Primary care physicians, endocrinologists, diabetes educators, dietitians, and mental health professionals
- **Support groups:** Local and online communities offering peer support and information
- **Educational organizations:** Such as the American Diabetes Association (ADA) and the International Diabetes Federation (IDF), providing resources, research updates, and advocacy
- **Mobile apps and technology:** Tools for blood sugar monitoring, medication reminders, and lifestyle tracking
- **Books and online materials:** Informative resources on diabetes management, recipes, and exercise plans. A ccessing these resources can help individuals effectively manage their diabetes and improve their quality of life.

Conclusion

Diabetes is a complex chronic condition that affects millions worldwide, characterized by elevated blood sugar levels due to either insufficient insulin production or ineffective use of insulin by the body. Understanding its types—Type 1, Type 2, gestational, and diabetes insipidus—reveals diverse causes and management approaches. Type 1 diabetes results from autoimmune destruction of insulin-producing cells, while Type 2 diabetes involves insulin resistance and declining pancreatic function, often linked to lifestyle factors like obesity and physical inactivity. Gestational diabetes arises during pregnancy and poses risks to both mother and child if left unmanaged. Diabetes insipidus, though less common, impacts fluid balance due to either brain or kidney dysfunction.

Managing diabetes requires a multifaceted approach. Early diagnosis through blood tests like fasting plasma glucose and A1C tests is critical, followed by personalized treatment plans encompassing diet, exercise, medication, and possibly insulin therapy. Continuous glucose monitoring and insulin pumps have transformed management, while medications like metformin and lifestyle adjustments help maintain stable blood sugar levels. Nutrition plays a pivotal role, emphasizing balanced meals rich in whole grains, lean proteins, and fiber to regulate glucose intake.

Physical activity not only aids in weight management but also enhances insulin sensitivity and cardiovascular health. Regular health checkups monitor complications such as neuropathy, retinopathy, and cardiovascular disease, ensuring timely intervention. Psychological support is crucial, as diabetes management can impact mental health due to stress and lifestyle adjustments.

Prevention strategies focus on healthy lifestyle choices, including smoking cessation and moderation of alcohol consumption. Educating patients and caregivers fosters empowerment and informed decision-making, promoting adherence to treatment plans and reducing the risk of complications.

In conclusion, while diabetes presents significant challenges, advancements in medical technology, coupled with proactive management strategies and comprehensive support systems, empower individuals to lead fulfilling lives while effectively managing their condition. Collaborative efforts among healthcare providers, patients, and support networks are essential in combating the global burden of diabetes and improving outcomes for those affected.

Bibliography

1. American Dental Association. (n.d.). *Oral health.* https://diabetes.org/health-wellness/keeping-your-mouth-healthy
2. American Heart Association. (2024). *Cholesterol and diabetes.* https://www.heart.org/en/health-topics/diabetes/diabetes-complications-and-risks/cholesterol-abnormalities—diabetes
3. American Diabetes Association. (n.d.). *Eye complications.* https://diabetes.org/about-diabetes/complications/eye-complications
4. American Diabetes Association. (n.d.). *Foot care tips.* https://diabetes.org/health-wellness/diabetes-and-your-feet/foot-care-tips#:~:text=Wash%20your%20feet%20well%20every,which%20can%20lead%20to%20infections
5. American Diabetes Association. (n.d.). *Diabetes technology guide.* https://diabetes.org/living-with-diabetes/treatment-care/diabetes-technology-guide
6. Government of Canada. (2024, January 12). *Tobacco and type 2 diabetes.* https://www.canada.ca/en/health-canada/services/health-concerns/tobacco/legislation/tobacco-product-labelling/tobacco-type-2-diabetes.html
7. American Veterinary Medical Association. (n.d.). *Diabetes in pets.* https://www.avma.org/resources-tools/pet-owners/petcare/diabetes-pets
8. Centers for Disease Control and Prevention. (2024, May 14). *Type 1 diabetes.* https://www.cdc.gov/diabetes/about/about-type-1-diabetes.html
9. Mayo Clinic. (2020, September 15). *Foot care with diabetes.* https://connect.mayoclinic.org/blog/take-charge-healthy-aging/newsfeed-post/foot-care-with-diabetes/
10. Mayo Clinic. (n.d.). *Gestational diabetes.* https://www.mayoclinic.org/diseases-conditions/gestational-diabetes/diagnosis-treatment/drc-20355345
11. Grossman, L. D., Roscoe, R., & Shack, A. R. (2018, April). Complementary and alternative medicine for diabetes. *Canadian Journal of Diabetes, 42*(Suppl 1), S154–S161. https://www.canadianjournalofdiabetes.com/article/S1499-2671(17)30833-X/fulltext
12. National Eye Institute. (2024, December 10). *Diabetic retinopathy.* https://www.nei.nih.gov/learn-about-eye-health/eye-conditions-and-diseases/diabetic-retinopathy#:~:text=Diabetic%20retinopathy%20is%20caused%20by,vessels%20all%20over%20the%20body
13. National Institute of Diabetes and Digestive and Kidney Diseases. (2021, April). *Diabetes, heart disease, and stroke.* https://www.niddk.nih.gov/health-information/diabetes/overview/preventing-problems/heart-disease-stroke#:~:text=Adults%20with%20diabetes%20are%20nearly,stroke%20as%20adults%20without%20diabetes.&text=The%20good%20news%20is%20that,having%20heart%20disease%20or%20stroke
14. National Institute of Diabetes and Digestive and Kidney Diseases. (2018, February). *What is diabetic neuropathy?* https://www.niddk.nih.gov/health-information/diabetes/overview/preventing-problems/nerve-damage-diabetic-neuropathies/what-is-diabetic-neuropathy

Living with Diabetes

CHAPTER 2

Introduction

Living with diabetes constitutes a lifelong journey marked by diligent self-care, informed decision-making, and ongoing adaptation to manage the complexities of this chronic condition effectively. Diabetes, whether Type 1, Type 2, or gestational, necessitates a multifaceted approach encompassing diet, exercise, medication adherence, and emotional well-being. This chapter delves into the fundamental aspects of living with diabetes, offering insights into practical strategies for maintaining blood sugar control, preventing complications, and optimizing overall health. By exploring these critical components, individuals with diabetes can empower themselves with the knowledge and tools needed to navigate daily challenges and lead fulfilling lives despite the demands of their condition.

1. How can someone newly diagnosed with diabetes cope with the diagnosis?

Coping with a new diabetes diagnosis involves educating oneself about the condition through reliable sources like healthcare providers and reputable organizations such as the American Diabetes Association. Seek support from healthcare professionals who can provide personalized guidance on diet, exercise, and medication. Joining support groups or connecting with others who have diabetes can offer emotional support and practical tips. Gradually implementing lifestyle changes, such as adopting a balanced diet and incorporating physical activity into daily routines, can empower individuals to manage their condition effectively.

2. What is the role of a diabetes management plan?

A diabetes management plan is crucial for maintaining blood sugar levels within target ranges and preventing long-term complications. It typically includes personalized goals for diet, exercise, medication, and regular monitoring of blood glucose levels. The plan may also incorporate strategies for managing stress, recognizing and responding to high or low blood sugar levels, and addressing other health concerns such as blood pressure and cholesterol levels. By following a structured management plan, individuals with diabetes can optimize their health outcomes and enhance their quality of life.

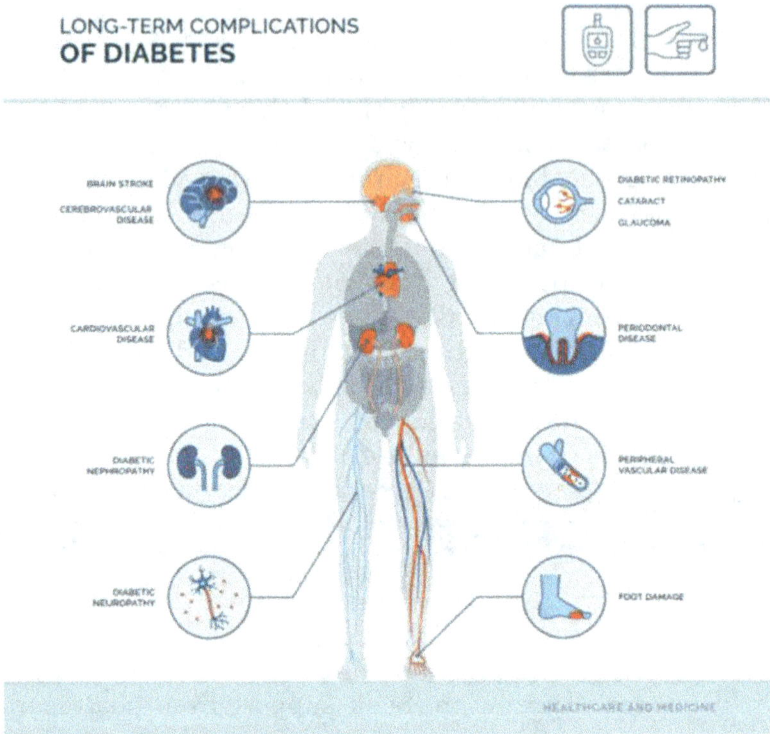

▲ Figure 2.1
Long-term complications of diabetes.

3. What are some common misconceptions about diabetes?

Common misconceptions about diabetes include beliefs that it is caused solely by consuming too much sugar, that only overweight individuals develop diabetes, or that diabetes is not a serious health condition. In reality, diabetes is a complex disorder influenced by genetic, lifestyle, and environmental factors. While excessive sugar intake can exacerbate diabetes, it is not the sole cause. Type 1 diabetes is an autoimmune condition that cannot be prevented through diet alone, and Type 2 diabetes can develop in individuals of any weight due to factors like genetics and physical inactivity. Understanding these misconceptions is crucial for promoting accurate information and dispelling stigma surrounding diabetes.

4. How can meal planning help in managing diabetes?

Meal planning plays a pivotal role in diabetes management by promoting balanced nutrition, regulating blood sugar levels, and supporting overall health. Planning meals and snacks in advance allows individuals to make informed choices about portion sizes, carbohydrate intake, and nutrient distribution throughout the day. Incorporating a variety of foods from different food groups—such as whole grains, lean proteins, fruits, vegetables, and healthy fats—helps to maintain steady blood glucose levels and provides essential vitamins and minerals. By consistently following a well-structured meal plan tailored to their dietary needs and preferences, individuals with diabetes can optimize their nutritional intake and enhance their ability to manage their condition effectively.

5. What are some healthy snacks for people with diabetes?

Healthy snacks for individuals with diabetes include options that provide sustained energy and help to regulate blood sugar levels. Examples include:

- **Nuts and seeds:** Almonds, walnuts, chia seeds, or pumpkin seeds are rich in healthy fats, protein, and fiber.
- **Yogurt:** Choose plain or Greek yogurt with no added sugars, and consider adding fresh berries or a sprinkle of nuts for extra flavor and nutrients.
- **Fruits:** Opt for whole fruits like apples, berries, or oranges, which provide natural sugars along with fiber and essential vitamins.
- **Vegetables:** Raw vegetables such as carrot sticks, cucumber slices, or bell pepper strips are low in calories and carbohydrates while offering fiber and antioxidants. These snacks can be combined with a source of protein or healthy fat to promote satiety and help maintain stable blood glucose levels between meals.

▲ Figure 2.2
Healthy snacks.

6. How does fiber benefit people with diabetes?

Fiber offers multiple health benefits for individuals with diabetes, including helping to regulate blood sugar levels, improve digestive health, and support weight management. Soluble fiber, found in foods like oats, beans, lentils, and fruits, forms a gel-like substance in the digestive tract that slows down the absorption of sugars and helps to prevent rapid spikes in blood glucose levels after meals. Insoluble fiber, found in whole grains, nuts, seeds, and vegetables, adds bulk to stool and promotes regular bowel movements, which can be beneficial for managing constipation—a common issue among individuals with diabetes. Including a variety of fiber-rich foods in the diet can contribute to overall health and help individuals with diabetes achieve better glycemic control over time.

7. What is carb counting and why is it important?

Carb counting, or carbohydrate counting, involves monitoring and managing the amount of carbohydrates consumed in meals and snacks to regulate blood sugar levels effectively. Carbohydrates have the most significant impact on blood glucose levels compared to fats and proteins, making it essential for individuals with diabetes to monitor their carbohydrate intake closely. By accurately counting carbohydrates and matching insulin doses or medication accordingly, individuals can better predict and control their blood sugar response to meals. This approach helps to prevent hyperglycemia (high blood sugar) and hypoglycemia (low blood sugar) episodes, optimize insulin therapy, and promote overall diabetes management. Learning carb counting skills empowers individuals with diabetes to make informed food choices, manage portion sizes, and maintain stable blood glucose levels throughout the day.

▲ **Figure 2.3**
Carb counting.

8. How can people with diabetes eat out at restaurants?

Eating out at restaurants can be managed effectively by individuals with diabetes with careful planning and mindful choices. Strategies include:

- Researching restaurant menus in advance to identify diabetes-friendly options and nutritional information when available.
- Choosing dishes that are grilled, baked, or steamed rather than fried, breaded, or heavily sauced.
- Opting for lean protein sources such as grilled chicken or fish, and selecting vegetables or salads as side dishes instead of high-carbohydrate options like fries or bread.
- Requesting modifications to meals, such as substituting whole grains for refined carbohydrates or asking for dressings and sauces on the side to control portions and reduce added sugars.
- Being mindful of portion sizes and considering sharing larger entrees or packing leftovers to avoid overeating.
- Monitoring blood sugar levels before and after meals, adjusting insulin doses or medications as needed, and staying hydrated with water or unsweetened beverages.

By applying these strategies, individuals with diabetes can enjoy dining out while making informed choices that support their health goals and help to manage blood sugar levels effectively.

9. What is the plate method for diabetes management?

The plate method is a visual tool used in diabetes management to guide meal planning and portion control. It involves dividing a standard dinner plate into sections to ensure a balanced and nutritious meal:

- Fill half of the plate with nonstarchy vegetables such as leafy greens, broccoli, peppers, or tomatoes, which are low in carbohydrates and calories but rich in vitamins, minerals, and fiber.
- Allocate one-quarter of the plate to lean protein sources like grilled chicken, turkey, fish, tofu, or beans, which provide essential amino acids for muscle repair and maintenance without contributing excessive amounts of fat or carbohydrates.

- Dedicate the remaining one-quarter of the plate to carbohydrate-containing foods such as whole grains (e.g., brown rice, quinoa, whole wheat pasta), starchy vegetables (e.g., potatoes, corn, peas), or legumes, which supply energy and essential nutrients while promoting satiety and supporting blood sugar control. Using the plate method helps individuals with diabetes visualize portion sizes, balance their nutrient intake, and create well-rounded meals that contribute to stable blood glucose levels and overall health. By adopting this approach to meal planning, individuals can simplify their dietary choices, make healthier food selections, and support effective diabetes management.

10. How can someone with diabetes maintain a healthy weight?

Maintaining a healthy weight is important for managing diabetes and reducing the risk of complications. Strategies include:

- **Following a balanced diet:** Consuming nutrient-dense foods, controlling portion sizes, and monitoring carbohydrate intake can help manage weight and blood sugar levels.
- **Regular physical activity:** Engaging in aerobic exercises such as walking, swimming, or cycling, as well as strength training exercises, can promote weight loss, improve insulin sensitivity, and enhance overall fitness.
- **Monitoring food intake:** Keeping track of meals and snacks, practicing mindful eating, and avoiding emotional eating can support weight management goals.
- **Setting realistic goals:** Establishing achievable targets for weight loss or weight maintenance, with the guidance of healthcare providers, can provide motivation and structure.
- **Seeking support:** Consulting with a registered dietitian, diabetes educator, or healthcare team for personalized advice and encouragement can facilitate successful weight management efforts. By incorporating these strategies into daily routines and making sustainable lifestyle changes, individuals with diabetes can achieve and maintain a healthy weight, improve

diabetes management outcomes, and enhance overall well-being.

11. What types of exercises are recommended for people with diabetes?

Exercise is beneficial for people with diabetes and should include a combination of aerobic, strength training, and flexibility exercises:

- Aerobic exercises such as brisk walking, jogging, cycling, or dancing help to improve cardiovascular health, increase insulin sensitivity, and lower blood glucose levels.
- Strength training exercises using resistance bands, free weights, or weight machines build muscle mass, enhance metabolism, and support weight management.
- Flexibility exercises like yoga or stretching routines improve joint mobility, reduce muscle tension, and promote relaxation. Engaging in regular physical activity, with the guidance of healthcare providers, helps to manage blood sugar levels, reduce the risk of cardiovascular complications, and enhance overall fitness and well-being. It is important for individuals with diabetes to choose activities they enjoy and can sustain over time, while monitoring blood sugar levels before, during, and after exercise to ensure safety and optimize health benefits.

12. How does stress impact diabetes management?

Stress can affect diabetes management by causing hormonal changes that raise blood glucose levels, impair insulin sensitivity, and interfere with self-care routines. Chronic stress may lead to emotional eating, poor sleep quality, and decreased physical activity, which can further exacerbate blood sugar fluctuations and diabetes-related complications. Managing stress effectively is essential for individuals with diabetes and can be achieved through various techniques such as:

- **Relaxation techniques:** Practicing deep breathing, progressive muscle relaxation, or mindfulness meditation promotes relaxation and reduces stress levels.

- **Physical activity:** Engaging in regular exercise releases endorphins, improves mood, and helps to alleviate stress and tension.
- **Support network:** Seeking emotional support from family, friends, or support groups provides encouragement, empathy, and practical coping strategies.
- **Time management:** Organizing daily tasks, prioritizing responsibilities, and setting realistic goals can reduce stress and improve time efficiency. By incorporating stress management techniques into daily routines and addressing sources of stress proactively, individuals with diabetes can enhance their ability to cope with the challenges of diabetes management, maintain stable blood sugar levels, and improve overall quality of life.

13. What are some stress management techniques for people with diabetes?

Stress management techniques for people with diabetes include:

- **Exercise**: Engaging in regular physical activity such as walking, jogging, or yoga helps to reduce stress hormones and promote relaxation.
- **Deep breathing:** Practicing deep breathing exercises, such as diaphragmatic breathing or guided imagery, can calm the mind and lower stress levels.
- **Meditation:** Mindfulness meditation involves focusing on the present moment without judgment, which can help to reduce anxiety and improve emotional well-being.
- **Progressive muscle relaxation:** This technique involves tensing and then relaxing different muscle groups in the body to release physical tension and reduce stress.
- **Hobby or leisure activities:** Engaging in enjoyable activities such as reading, gardening, or listening to music provides a mental distraction from stressors and promotes relaxation.

- **Support network:** Seeking emotional support from family, friends, or support groups provides encouragement, empathy, and opportunities to share experiences and coping strategies. By incorporating these stress management techniques into daily routines, individuals with diabetes can reduce stress levels, enhance overall well-being, and improve their ability to manage blood sugar levels effectively.

14. What are the benefits of regular blood sugar monitoring?

Regular blood sugar monitoring provides valuable information for individuals with diabetes to assess their blood glucose levels, evaluate the effectiveness of diabetes management strategies, and make informed decisions about diet, exercise, and medication. Key benefits include:

- **Maintaining target ranges:** Monitoring blood sugar levels helps individuals to achieve and maintain target ranges recommended by healthcare providers for optimal diabetes management.
- **Preventing complications:** By detecting and addressing high or low blood sugar levels promptly, individuals can reduce the risk of short-term complications such as hyperglycemia (high blood sugar) or hypoglycemia (low blood sugar).
- **Informing treatment adjustments:** Blood sugar data guides adjustments to insulin doses, oral medications, meal plans, or exercise routines to optimize blood glucose control and prevent fluctuations.
- **Promoting self-awareness:** Monitoring blood sugar levels fosters awareness of how diet, physical activity, stress, illness, or medications impact blood glucose levels, empowering individuals to make proactive choices.
- **Enhancing communication with healthcare providers:** Sharing blood sugar records during healthcare visits facilitates productive discussions with healthcare providers about diabetes management progress, challenges, and adjustments. By integrating regular blood sugar monitoring into daily self-care routines and utilizing technology such as glucose meters or continuous glucose monitors (CGMs), individuals

with diabetes can enhance their ability to manage their condition effectively, reduce the risk of complications, and improve overall quality of life.

15. What is continuous glucose monitoring (CGM)?

Continuous glucose monitoring (CGM) is a technology that tracks blood sugar levels continuously throughout the day and night, providing real-time data to individuals with diabetes or their caregivers. A CGM system typically consists of a small sensor placed under the skin to measure glucose levels in the interstitial fluid, a transmitter that sends data wirelessly to a receiver or smartphone app, and software that displays glucose trends, alerts for high or low glucose levels, and patterns over time. Key benefits of CGM include:

- **Real-time data:** CGM systems provide immediate feedback on blood sugar levels, allowing individuals to make timely adjustments to diet, exercise, or medication to maintain target ranges.
- **Trend analysis:** CGM software displays glucose trends, patterns, and variability over hours and days, offering insights into factors affecting blood sugar control.
- **Alerts and alarms:** CGM alerts notify users of high or low glucose levels, impending hypoglycemia or hyperglycemia, or rapid changes in glucose levels, prompting proactive intervention.
- **Remote monitoring:** Some CGM systems offer remote sharing of glucose data with healthcare providers or caregivers, facilitating collaborative diabetes management and timely intervention.
- **Historical data review:** CGM software stores historical glucose data that can be reviewed during healthcare visits to evaluate diabetes management progress, identify trends, and adjust treatment plans.

By using CGM technology as part of comprehensive diabetes management, individuals can gain greater insight into their blood sugar levels, improve glycemic control, reduce the risk of complications, and enhance their overall quality of life.

16. How does insulin therapy work?

Insulin therapy is a cornerstone of treatment for individuals with Type 1 diabetes and may be prescribed for some individuals with Type 2 diabetes when lifestyle changes and oral medications are insufficient to achieve target blood sugar levels. Insulin is a hormone produced by the pancreas that regulates glucose metabolism by enabling cells to absorb glucose from the bloodstream for energy or storage. In diabetes, either the pancreas produces insufficient insulin (Type 1 diabetes) or the body's cells become resistant to insulin's effects (Type 2 diabetes), leading to elevated blood sugar levels. Insulin therapy involves administering exogenous (external) insulin to replace or supplement the body's natural insulin production, thereby lowering blood glucose levels and preventing hyperglycemia (high blood sugar). Types of insulin vary in onset (how quickly they begin to work), peak (when they have the strongest effect), and duration (how long they continue to lower blood sugar levels), and may include:

- **Rapid-acting insulin:** Begins to work within 15 minutes, peaks in 1 to 2 hours, and lasts for 3 to 4 hours.
- **Short-acting insulin:** Begins to work within 30 minutes to 1 hour, peaks in 2 to 3 hours, and lasts for 3 to 6 hours.
- **Intermediate-acting insulin:** Begins to work within 1 to 2 hours, peaks in 4 to 12 hours, and lasts for 12 to 18 hours.
- **Long-acting insulin:** Begins to work several hours after injection, has no pronounced peak, and lasts for up to 24 hours or longer. Insulin therapy is administered via injection using insulin syringes, pens, or insulin pumps that deliver insulin continuously. The dosage and timing of insulin administration are individualized based on factors such as blood sugar levels, meal plans, physical activity, and personal preferences. By achieving and maintaining optimal blood sugar control through insulin therapy, individuals with diabetes can reduce the risk of complications, enhance quality of life, and promote long-term health.

17. What are some common side effects of diabetes medications?

Common side effects of diabetes medications may vary depending on the type of medication prescribed and individual response, but can include:

- **Hypoglycemia (low blood sugar):** Some diabetes medications, such as insulin or sulfonylureas, may lower blood sugar levels excessively, leading to symptoms such as shakiness, sweating, confusion, dizziness, or fainting.
- **Weight gain:** Certain diabetes medications, including insulin, thiazolidinediones (TZDs), or sulfonylureas, may contribute to weight gain due to their effects on metabolism or appetite regulation.
- **Gastrointestinal issues:** Digestive side effects such as nausea, vomiting, diarrhea, or abdominal discomfort may occur with medications like metformin or GLP-1 receptor agonists.
- **Allergic reactions:** Some individuals may experience allergic reactions, such as skin rash, itching, swelling, or difficulty breathing, in response to certain diabetes medications.
- **Liver or kidney complications:** In rare cases, diabetes medications may affect liver or kidney function, particularly with prolonged use or in individuals with pre-existing conditions.
- **Fluid retention:** Medications like TZDs may cause fluid retention or swelling in the legs, ankles, or hands, which should be monitored and reported to healthcare providers.
- **Bone health:** Certain diabetes medications, such as TZDs, may increase the risk of bone fractures or affect bone density, especially with long-term use.

It is essential for individuals with diabetes to be aware of potential side effects associated with their prescribed medications, monitor for symptoms, and communicate any concerns or changes in health to healthcare providers promptly. Healthcare providers can adjust medication regimens, recommend alternative treatments, or provide supportive care to manage side effects effectively while optimizing diabetes management outcomes.

18. How can people with diabetes manage their medications effectively?

Effectively managing diabetes medications involves several key strategies to ensure safety, adherence, and optimal therapeutic outcomes:

- **Understanding medication instructions:** Educate oneself about prescribed medications, including dosage, timing of administration, potential side effects, and interactions with food or other medications.
- **Establishing a medication routine:** Develop a consistent schedule for taking medications, incorporating them into daily routines or using reminders such as alarms or pill organizers to enhance adherence.
- **Monitoring blood sugar levels:** Regularly monitor blood glucose levels as directed by healthcare providers to assess the effectiveness of medications and make necessary adjustments.
- **Communicating with healthcare providers:** Maintain open communication with healthcare providers about medication adherence, any changes in health or symptoms, and concerns regarding side effects or treatment effectiveness.
- **Reviewing medication lists:** Periodically review medication lists with healthcare providers to ensure they are up to date, appropriate for individual health needs, and effectively managing diabetes.
- **Storing medications properly:** Store medications according to package instructions, such as avoiding exposure to extreme temperatures or direct sunlight, to maintain potency and safety.
- **Seeking support:** Engage with healthcare providers, pharmacists, or diabetes educators for guidance on medication management, including proper administration techniques and strategies for minimizing side effects. By actively participating in medication management, individuals with diabetes can optimize treatment outcomes, reduce the risk of complications, and promote overall health and well-being.

19. What is the importance of regular eye exams for people with diabetes?

Regular eye exams are critical for individuals with diabetes to detect and manage eye-related complications, such as diabetic retinopathy, macular edema, or cataracts, which can affect vision and eye health. Diabetes can increase the risk of developing these conditions due to fluctuating blood sugar levels and changes in blood vessels within the retina, leading to potential vision loss if left untreated. Key benefits of regular eye exams include:

- **Early detection:** Routine eye exams can identify signs of diabetic eye disease before symptoms occur, allowing for prompt intervention and treatment to preserve vision.
- **Monitoring eye health:** Comprehensive eye exams assess visual acuity, peripheral vision, intraocular pressure, and the health of the retina, optic nerve, and other eye structures.
- **Preventing complications:** Timely treatment options, such as laser therapy, injections, or surgery, can prevent or slow the progression of diabetic eye disease and minimize visual impairment.
- **Educating individuals:** Eye care professionals provide education on maintaining eye health, managing diabetes-related eye complications, and adhering to recommended treatment plans.
- **Coordinating care:** Collaboration between healthcare providers, including primary care physicians, endocrinologists, and eye specialists, ensures comprehensive diabetes management and integrated treatment of eye-related concerns. Individuals with diabetes should schedule regular eye exams as recommended by healthcare providers, adhere to prescribed treatment plans, and promptly report any changes in vision or eye health. By prioritizing eye care and early intervention, individuals can preserve vision, reduce the impact of diabetic eye complications, and maintain quality of life.

20. How can people with diabetes protect their kidneys?

Protecting kidney health is crucial for individuals with diabetes, as diabetes-related kidney disease (diabetic nephropathy) can lead to kidney failure if left untreated. Strategies to protect kidneys include:

- **Controlling blood sugar levels:** Maintain target blood glucose levels recommended by healthcare providers through diet, exercise, medication adherence, and regular monitoring.
- **Managing blood pressure:** Monitor and control blood pressure levels to reduce strain on the kidneys and slow the progression of diabetic kidney disease (nephropathy).
- **Monitoring kidney function:** Regularly undergo kidney function tests, such as urine albumin-to-creatinine ratio (ACR) and estimated glomerular filtration rate (eGFR), to assess kidney health and detect early signs of dysfunction.
- **Avoiding nephrotoxic substances:** Minimize or avoid the use of medications, substances, or recreational drugs that can harm kidney function, particularly in individuals with pre-existing kidney disease.
- **Maintaining a healthy lifestyle:** Adopt a balanced diet low in sodium and saturated fats, engage in regular physical activity, avoid smoking, and limit alcohol consumption to support overall kidney health.
- **Seeking early intervention:** Consult healthcare providers promptly if experiencing symptoms such as changes in urinary habits, swelling in the legs or ankles, fatigue, or unexplained weight loss. By implementing these kidney-protective strategies and collaborating with healthcare providers to monitor kidney function regularly, individuals with diabetes can reduce the risk of diabetic nephropathy, preserve kidney function, and promote long-term health and well-being.

21. What is diabetic neuropathy and how is it managed?

Diabetic neuropathy is a type of nerve damage that can develop in individuals with diabetes, particularly when blood sugar levels are consistently high over time. Neuropathy may affect nerves throughout the body, causing symptoms such as pain, tingling, numbness, or weakness in the extremities (peripheral neuropathy), digestive issues (autonomic neuropathy), or sexual dysfunction (sexual neuropathy). Management strategies for diabetic neuropathy include:

- **Blood sugar control:** Maintaining target blood glucose levels recommended by healthcare providers can help to prevent or slow the progression of nerve damage.
- **Medication management:** Prescribed medications, such as antidepressants, anticonvulsants, or topical treatments, may be used to manage neuropathic pain or discomfort.
- **Pain management techniques:** Applying heat or cold therapy, practicing relaxation techniques, using supportive footwear, or engaging in physical therapy can alleviate neuropathic symptoms and improve quality of life.
- **Foot care:** Proper foot hygiene, regular foot inspections, wearing comfortable shoes, and avoiding injury or pressure to the feet can prevent foot ulcers or infections associated with peripheral neuropathy.
- **Lifestyle modifications:** Adopting a balanced diet, engaging in regular physical activity, managing stress effectively, and avoiding alcohol or tobacco use support overall nerve health and diabetes management.
- **Regular monitoring:** Routine assessments by healthcare providers to evaluate nerve function, assess symptoms, and adjust treatment plans as needed ensure comprehensive management of diabetic neuropathy. By addressing diabetic neuropathy with a multidisciplinary approach that includes blood sugar control, symptom management, preventive care, and lifestyle adjustments, individuals with diabetes can minimize the impact of nerve damage, improve quality of life, and maintain optimal health outcomes.

22. How can people with diabetes take care of their skin?

Skin care is important for individuals with diabetes to prevent complications such as dry skin, infections, or slow wound healing, which can occur due to fluctuations in blood sugar levels or impaired circulation. Skin care practices include:

- **Daily cleansing:** Wash skin gently with mild soap and lukewarm water, focusing on areas prone to sweat or moisture buildup, such as underarms, groin, and between toes.
- **Moisturizing:** Apply a fragrance-free moisturizer or emollient to keep skin hydrated, particularly after bathing or when skin feels dry, to prevent itching or cracking.
- **Inspecting skin:** Check skin regularly for cuts, scratches, blisters, or changes in color, and consult healthcare providers promptly if noticing any signs of infection or slow healing.
- **Protecting feet:** Wear properly fitting shoes and socks to prevent blisters or pressure sores, avoid walking barefoot, and maintain good foot hygiene to reduce the risk of foot ulcers.
- **Managing wounds:** Clean cuts or scrapes with mild soap and water, apply an antiseptic or antibiotic ointment, cover with a sterile bandage, and seek medical attention for deep or infected wounds.
- **Avoiding harsh products:** Use gentle skincare products without alcohol, fragrances, or harsh chemicals that can irritate or dry out the skin, and test new products on a small area before widespread use. By incorporating these skin care practices into daily routines and seeking prompt medical attention for any skin-related concerns, individuals with diabetes can maintain healthy skin, reduce the risk of complications, and support overall well-being.

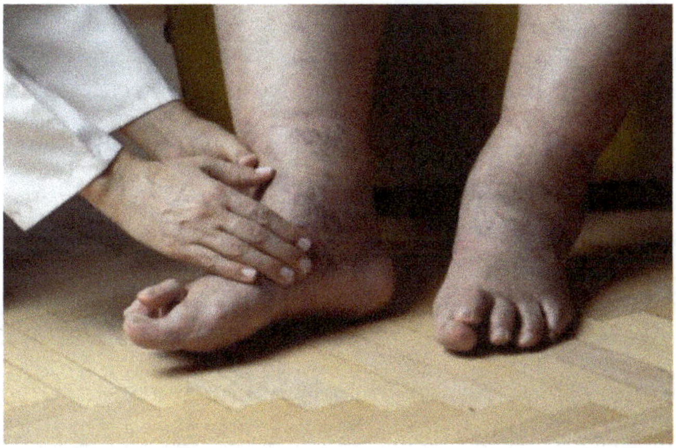

▲ Figure 2.4
Swelling in the feet caused by diabetes.

23. What are the signs of a diabetes emergency?

Signs of a diabetes emergency include acute complications such as severe hypoglycemia, diabetic ketoacidosis (DKA), or hyperosmolar hyperglycemic state (HHS), which require immediate medical attention. Recognizing and responding to these emergencies promptly can prevent serious complications and potentially life-threatening situations. Key signs include:

- **Severe hypoglycemia:** Symptoms may include confusion, dizziness, sweating, shakiness, irritability, weakness, or loss of consciousness due to dangerously low blood sugar levels.
- **Diabetic ketoacidosis (DKA):** Symptoms may include excessive thirst, frequent urination, abdominal pain, nausea, vomiting, fruity breath odor, rapid breathing, or confusion due to high blood sugar levels and ketone buildup in the blood.
- **Hyperosmolar hyperglycemic state (HHS):** Symptoms may include extreme thirst, dry mouth, confusion, fever, warm or dry skin, rapid heartbeat, or unconsciousness due to very high blood sugar levels and severe dehydration. In any diabetes emergency, immediate action is crucial. Individuals experiencing severe symptoms should seek medical assistance or call emergency services (such as 911 in the United States) for prompt evaluation, treatment, and stabilization. Healthcare providers can assess the underlying cause

of the emergency, administer appropriate interventions, and provide supportive care to resolve symptoms, restore metabolic balance, and prevent recurrence. Education on recognizing emergency signs, developing an emergency action plan, and communicating emergency contact information with caregivers or loved ones can empower individuals with diabetes to respond effectively to emergencies and prioritize their health and safety.

24. How can people with diabetes prepare for travel?

Planning and preparation are essential for individuals with diabetes to manage their condition effectively while traveling, whether for short trips or extended vacations. Key strategies include:

- **Consult healthcare providers:** Schedule a pretravel consultation with healthcare providers to review diabetes management plans, update prescriptions, and obtain necessary medications, supplies, or immunizations.
- **Pack diabetes supplies:** Pack sufficient quantities of insulin, oral medications, glucose meters, test strips, lancets, syringes or insulin pens, snacks, and any other supplies needed for daily diabetes care.
- **Carry medical identification:** Wear or carry a medical alert bracelet or necklace indicating diabetes, along with emergency contact information and healthcare provider details in case of unforeseen medical situations.
- **Plan meals and snacks:** Research food options available at travel destinations, pack healthy snacks such as fruit, nuts, or granola bars, and consider carrying portable meal options to maintain balanced nutrition and blood sugar control.
- **Manage time zone changes:** Adjust insulin dosing or medication schedules as needed to accommodate time zone changes during travel and consult healthcare providers for guidance on timing adjustments.
- **Monitor blood sugar levels:** Test blood glucose levels regularly, especially during travel, time zone changes, or changes in routine, and maintain awareness of symptoms indicating hypo- or hyperglycemia.

- **Stay hydrated and active:** Drink plenty of water, engage in regular physical activity such as walking or stretching during travel, and take breaks to prevent blood circulation issues or discomfort.
- **Be prepared for emergencies:** Carry a diabetes emergency kit with extra supplies, snacks, fast-acting glucose, and contact information for healthcare providers or local medical services in case of unexpected health concerns. By proactively planning for diabetes management during travel, maintaining open communication with healthcare providers, and being prepared for unforeseen circumstances, individuals with diabetes can enjoy safe and enjoyable travel experiences while prioritizing their health and well-being.

25. What should people with diabetes consider when dining out?

Dining out can present challenges for individuals with diabetes in managing blood sugar levels and making healthy food choices. Strategies for dining out include:

- **Plan ahead:** Review restaurant menus online, consider portion sizes, and choose venues that offer a variety of balanced meal options, including lean proteins, vegetables, whole grains, and fruits.
- **Modify menu selections:** Request modifications such as substituting steamed or grilled options for fried foods, asking for dressings or sauces on the side, or choosing smaller portion sizes to control carbohydrate intake.
- **Monitor portion sizes:** Be mindful of portion sizes served at restaurants, consider sharing meals or packing leftovers for later, and avoid overeating to maintain blood sugar control and prevent post meal spikes.
- **Monitor carbohydrate intake:** Estimate carbohydrate content in restaurant meals, focus on whole grains, fruits, and vegetables, and balance carbohydrate intake with protein and healthy fats to minimize blood sugar fluctuations.
- **Choose beverages wisely:** Opt for water, unsweetened tea, or other low-calorie beverages instead

of sugary drinks or alcoholic beverages that can impact blood sugar levels and overall health.

- **Carry diabetes supplies:** Bring glucose meters, test strips, fast-acting glucose, insulin, or medications as needed, and plan for adjustments in insulin dosing or medication schedules based on meal choices.
- **Communicate with restaurant staff**: Inform servers or chefs about dietary preferences, food allergies, or special requests related to diabetes management to ensure meal preparation aligns with individual health needs. By applying these strategies when dining out, individuals with diabetes can make informed choices, maintain blood sugar control, and enjoy dining experiences while supporting overall health and well-being.

26. How can people with diabetes manage stress effectively?

Managing stress effectively is important for individuals with diabetes to support overall well-being and optimize diabetes management. Strategies for stress management include:

- **Relaxation techniques:** Practice deep breathing, progressive muscle relaxation, guided imagery, or mindfulness meditation to promote relaxation, reduce stress hormones, and improve emotional well-being.
- **Physical activity:** Engage in regular exercise such as walking, jogging, swimming, or yoga to release endorphins, improve mood, and reduce stress levels.
- **Healthy lifestyle habits:** Maintain a balanced diet, prioritize adequate sleep, limit caffeine or alcohol intake, and avoid smoking to support physical health and resilience against stress.
- **Time management:** Organize daily routines, prioritize tasks, set realistic goals, and delegate responsibilities when possible to reduce stress related to time pressures or deadlines.
- **Social support:** Seek emotional support from family members, friends, or support groups, share experiences, and receive encouragement, empathy, and practical coping strategies.

- **Professional support:** Consult healthcare providers, counselors, therapists, or diabetes educators for guidance on stress management techniques, coping skills, and mental health support. By incorporating stress management strategies into daily routines, individuals with diabetes can reduce stress levels, enhance resilience, improve blood sugar control, and promote overall health and quality of life.

27. What are some stress management techniques for people with diabetes?

Stress management techniques for people with diabetes include:

- **Exercise:** Engaging in regular physical activity such as walking, jogging, or yoga helps to reduce stress hormones and promote relaxation.
- **Deep breathing:** Practicing deep breathing exercises, such as diaphragmatic breathing or guided imagery, can calm the mind and lower stress levels.
- **Meditation:** Mindfulness meditation involves focusing on the present moment without judgment, which can help to reduce anxiety and improve emotional well-being.
- **Progressive muscle relaxation:** This technique involves tensing and then relaxing different muscle groups in the body to release physical tension and reduce stress.
- **Hobby or leisure activities:** Engaging in enjoyable activities such as reading, gardening, or listening to music provides a mental distraction from stressors and promotes relaxation.
- **Support network:** Seeking emotional support from family, friends, or support groups provides encouragement, empathy, and opportunities to share experiences and coping strategies. By incorporating these stress management techniques into daily routines, individuals with diabetes can reduce stress levels, enhance overall well-being, and improve their ability to manage blood sugar levels effectively.

28. How can people with diabetes stay motivated to manage their condition?

Managing diabetes requires ongoing motivation and commitment. Here are some effective strategies:

- **Setting realistic goals:** Establish achievable short-term and long-term goals related to blood sugar control, diet, exercise, and overall health.
- **Tracking progress:** Monitor blood sugar levels, weight, physical activity, and adherence to medication or insulin regimens. Tracking progress can provide a sense of accomplishment and motivation.
- **Seeking support:** Connect with healthcare providers, diabetes educators, and support groups. Sharing experiences and receiving encouragement from others facing similar challenges can boost motivation and provide valuable insights.
- **Educating oneself:** Stay informed about diabetes management strategies, new treatment options, and lifestyle recommendations. Knowledge empowers individuals to make informed decisions and take proactive steps in their care.

29. What is the role of a registered dietitian in diabetes management?

A registered dietitian (RD) plays a crucial role in diabetes management by providing personalized nutrition guidance and support:

- **Personalized meal planning:** RDs assess individual nutritional needs based on factors like age, weight, activity level, and diabetes treatment plan. They create tailored meal plans that emphasize balanced nutrition, portion control, and carbohydrate management to help stabilize blood sugar levels.
- **Nutrition education:** RDs educate individuals about the impact of food choices on blood glucose levels and overall health. They provide practical strategies for incorporating healthy eating habits into daily routines, such as reading food labels, making healthier substitutions, and dining out sensibly.

- **Behavioral counseling:** RDs offer behavioral counseling to address barriers to dietary adherence, promote lifestyle changes, and support long-term dietary modifications. They empower individuals with skills and knowledge to manage diabetes through sustainable dietary practices.
- **Collaboration with healthcare team:** RDs collaborate with healthcare providers, including physicians, nurses, and diabetes educators, to integrate nutrition therapy into comprehensive diabetes care plans. They monitor progress, adjust meal plans as needed, and provide ongoing support to optimize health outcomes.

30. How can people with diabetes handle special occasions and holidays?

Managing diabetes during special occasions and holidays requires planning and mindful decision-making to maintain blood sugar control:

- **Plan ahead:** Anticipate meal options and potential challenges. Consider eating smaller portions throughout the day to balance carbohydrate intake.
- **Make healthy food choices:** Opt for lean proteins, vegetables, and whole grains. Limit high-sugar or high-fat foods, and enjoy sweets or treats in moderation while considering their impact on blood sugar levels.
- **Monitor blood sugar levels:** Check blood glucose levels frequently, especially after meals or snacks. Adjust insulin doses or medication as advised by healthcare providers to maintain target ranges.
- **Stay active:** Incorporate physical activity into holiday celebrations, such as taking walks, dancing, or participating in outdoor activities. Exercise helps regulate blood sugar levels and promotes overall well-being.
- **Communicate with healthcare team:** Discuss holiday plans with healthcare providers in advance. Seek guidance on adjusting medication schedules, managing stress, and handling unexpected situations that may affect diabetes management. By proactively planning and making informed choices, individuals with

diabetes can enjoy special occasions and holidays while prioritizing their health and well-being.

31. What are some common challenges in diabetes management?

Diabetes management involves navigating various challenges that impact daily life and overall health:

- **Maintaining a healthy diet:** Balancing carbohydrate intake, portion sizes, and meal timing to stabilize blood sugar levels can be challenging.
- **Regular exercise:** Finding time for physical activity, overcoming barriers such as weather or fatigue, and adjusting exercise routines based on blood sugar levels and medication regimens.
- **Medication adherence:** Remembering to take medications or insulin as prescribed, managing side effects, and following up with healthcare providers for adjustments.
- **Managing stress:** Stress can affect blood sugar levels and overall well-being. Finding effective stress management techniques, such as relaxation exercises or hobbies, is important.
- **Routine healthcare visits:** Scheduling and attending regular checkups, eye exams, and screenings to monitor diabetes-related complications and adjust treatment plans.
- **Support system:** Building a support network, seeking guidance from healthcare providers, and connecting with peers in diabetes support groups for encouragement and practical advice. By addressing these challenges with proactive strategies, individuals with diabetes can enhance their ability to manage the condition effectively and improve overall quality of life.

32. How can technology assist in diabetes management?

Technological advancements offer valuable tools and resources to support diabetes management efforts:

- **Glucose meters:** Devices for monitoring blood sugar levels at home, providing immediate feedback for adjusting diet, exercise, or medication.

- **Continuous glucose monitors (CGMs):** Wearable devices that track glucose levels throughout the day and night, offering real-time data and trends to prevent highs and lows.
- **Insulin pumps**: Automated devices for delivering insulin based on programmed settings, offering flexibility in insulin administration and enhancing control over blood sugar levels.
- **Mobile apps:** Applications for smartphones or tablets that track food intake, physical activity, medication adherence, and glucose levels. They provide personalized insights, reminders, and data sharing with healthcare providers.
- **Telemedicine:** Virtual platforms for remote consultations with healthcare providers, facilitating ongoing monitoring, education, and adjustments to diabetes management plans.
- **Educational resources:** Online platforms, webinars, and digital tools offering educational content, self-management tips, and community support for individuals with diabetes. By integrating technology into diabetes management routines, individuals can gain greater insight into their health metrics, enhance communication with healthcare providers, and make informed decisions to achieve optimal diabetes control and overall well-being.

33. What is the impact of diabetes on mental health?

Diabetes can significantly impact mental health due to the demands of managing a chronic condition and potential complications:

- **Psychological distress:** The constant need for self-care, monitoring blood sugar levels, and adhering to treatment regimens can lead to anxiety, depression, or feelings of burnout.
- **Diabetes distress:** Specific emotional responses to the challenges of living with diabetes, including fear of complications, frustration with dietary restrictions, or stress related to medication management.

- **Impact on quality of life:** Diabetes-related stressors may affect relationships, work productivity, sleep patterns, and overall well-being.
- **Coping strategies:** Individuals may develop coping strategies, such as seeking support from healthcare providers, participating in diabetes support groups, or engaging in stress-reduction activities like mindfulness or exercise.
- **Integrated care:** Collaborative efforts between healthcare providers and mental health professionals to address both physical and emotional aspects of diabetes management. By recognizing the impact of diabetes on mental health and implementing strategies to support emotional well-being, individuals can enhance their ability to cope with challenges, improve diabetes management outcomes, and maintain a positive outlook on life.

34. How can people with diabetes manage their emotional well-being?

Managing emotional well-being is essential for individuals with diabetes to cope with the challenges of living with a chronic condition:

- **Seeking support:** Connect with mental health professionals, counselors, or therapists who specialize in diabetes-related issues. Discuss emotional concerns, stressors, and coping strategies.
- **Joining support groups:** Participate in diabetes support groups, either in-person or online, to share experiences, receive encouragement, and learn practical tips from peers facing similar challenges.
- **Practicing stress-reduction techniques:** Incorporate relaxation techniques such as deep breathing, progressive muscle relaxation, guided imagery, or mindfulness meditation into daily routines to reduce stress levels.
- **Engaging in physical activity:** Regular exercise releases endorphins, improves mood, and promotes overall well-being. Choose activities that are enjoyable and fit individual fitness levels and preferences.

- **Maintaining routines:** Establish consistent daily routines for meal times, medication schedules, sleep patterns, and self-care practices to promote stability and reduce anxiety.
- **Setting realistic goals:** Break down larger tasks into smaller, manageable goals related to diabetes management, lifestyle changes, or personal interests. Celebrate achievements to boost self-confidence and motivation.
- **Balancing priorities:** Prioritize self-care, hobbies, social connections, and relaxation to maintain a balanced lifestyle that supports emotional and physical health. By prioritizing emotional well-being and integrating supportive strategies into daily life, individuals with diabetes can enhance resilience, cope with stressors effectively, and improve overall quality of life.

35. What are the benefits of joining a diabetes support group?

Diabetes support groups offer numerous benefits for individuals managing the challenges of diabetes:

- **Sense of community:** Connect with peers who understand the daily struggles, successes, and emotions associated with living with diabetes. Share experiences, exchange practical tips, and provide mutual support.
- **Emotional support:** Receive empathy, encouragement, and validation from group members facing similar challenges. Discuss concerns, fears, or frustrations related to diabetes management in a supportive environment.
- **Practical advice:** Gain insights into diabetes management strategies, medication adherence, blood sugar monitoring techniques, meal planning, and lifestyle adjustments from group discussions and shared experiences.
- **Education and empowerment:** Attend guest speaker presentations, educational workshops, or informational sessions on diabetes-related topics to enhance knowledge, skills, and self-confidence in managing diabetes.

- **Coping strategies:** Learn effective stress management techniques, relaxation exercises, mindfulness practices, or behavior modification strategies to improve coping skills and emotional resilience.
- **Motivation and accountability:** Set goals, track progress, and celebrate achievements with the encouragement and accountability of group members. Stay motivated to maintain healthy habits and prioritize diabetes self-care.
- **Access to resources:** Receive recommendations for reputable healthcare providers, diabetes educators, mental health professionals, or community resources that support comprehensive diabetes management. By participating in diabetes support groups, individuals can benefit from a sense of camaraderie, emotional support, practical advice, education, motivation, and resources that empower them to navigate the challenges of diabetes with confidence and resilience.

36. How can people with diabetes manage their condition during emergencies?

Preparation and planning are essential for individuals with diabetes to effectively manage their condition during emergencies:

- **Develop an emergency plan:** Create a personalized emergency plan in collaboration with healthcare providers. Include contact information for healthcare providers, emergency contacts, and relevant medical history details.
- **Keep supplies handy:** Maintain a well-stocked diabetes emergency kit containing essential supplies such as glucose meters, test strips, fast-acting glucose (e.g., glucose tablets or gel), insulin, syringes or insulin pens, medications, and medical identification.
- **Monitor blood sugar levels:** Regularly check blood glucose levels during emergencies or stressful situations. Be prepared to adjust insulin doses or medication schedules based on blood sugar readings and healthcare provider recommendations.
- **Stay hydrated and nourished:** Drink plenty of water to prevent dehydration and maintain hydration

levels. Have non perishable snacks or glucose-rich foods available to treat hypoglycemia or prevent low blood sugar levels.

- **Communicate with healthcare providers:** Inform healthcare providers about emergency situations, changes in health status, or symptoms that require medical attention. Follow their guidance on managing diabetes medications, monitoring blood sugar levels, and seeking medical care if needed.
- **Educate family and friends:** Ensure that family members, caregivers, or close contacts are familiar with your diabetes emergency plan, know how to administer emergency treatments, and understand when to seek medical assistance.
- **Seek medical attention:** In case of severe hypoglycemia, diabetic ketoacidosis (DKA), hyperosmolar hyperglycemic state (HHS), or other diabetes-related emergencies, seek immediate medical attention or call emergency services for prompt treatment and intervention. By proactively preparing for emergencies, staying informed, and following established diabetes management protocols, individuals can effectively manage their condition, minimize complications, and ensure their safety and well-being during unforeseen situations.

37. What should people with diabetes know about health insurance?

Understanding health insurance coverage is crucial for individuals with diabetes to access necessary medications, supplies, and healthcare services:

- **Coverage for diabetes supplies:** Review health insurance policies to understand coverage for glucose meters, test strips, lancets, insulin, oral medications, insulin pumps, continuous glucose monitors (CGMs), and other diabetes-related supplies.
- **Medication coverage:** Verify coverage for prescription medications used to manage diabetes, including insulin analogs, oral antidiabetic drugs, or injectable therapies. Understand copayment or coinsurance

amounts, formulary restrictions, and prior authorization requirements.

- **Access to healthcare providers:** Confirm network participation of primary care physicians, endocrinologists, diabetes educators, and specialists who provide comprehensive diabetes care. Understand referral processes and requirements for seeing out-of-network providers if needed.
- **Annual deductibles and out-of-pocket costs:** Determine annual deductibles, copayments, coinsurance percentages, and maximum out-of-pocket expenses under the health insurance plan. Budget for healthcare expenses and seek financial assistance or resources if facing financial challenges.
- **Preventive care services:** Take advantage of covered preventive care services, such as annual wellness visits, diabetes screenings, eye exams, foot exams, and laboratory tests to monitor diabetes-related health indicators and detect complications early.
- **Emergency coverage:** Understand coverage for emergency medical services, hospitalization, urgent care visits, and ambulance transportation in case of diabetes-related emergencies, severe hypoglycemia, or acute complications like diabetic ketoacidosis (DKA).
- **Appeals and grievances:** Familiarize yourself with the health insurance plan's appeals process for disputing coverage denials, requesting reconsideration of denied claims, or filing grievances related to denied services or benefits.
- **Patient rights and protections:** Know your rights under health insurance laws, including protections against coverage discrimination based on pre-existing conditions like diabetes. Advocate for fair treatment and access to medically necessary treatments and services. By staying informed about health insurance coverage, understanding policy terms and benefits, and advocating for comprehensive diabetes care, individuals can navigate insurance challenges, access essential healthcare resources, and effectively manage their diabetes while maintaining financial stability and peace of mind.

38. How can people with diabetes advocate for themselves in healthcare settings?

Advocating for oneself in healthcare settings empowers individuals with diabetes to receive personalized care, address concerns, and achieve optimal health outcomes:

- **Be informed:** Educate yourself about diabetes management, treatment options, and available resources. Stay updated on advancements in diabetes care, research findings, and guidelines from reputable healthcare organizations.
- **Ask questions:** Prepare a list of questions or concerns to discuss during healthcare appointments. Inquire about diabetes medications, treatment plans, blood sugar monitoring techniques, lifestyle modifications, and strategies for managing diabetes-related complications.
- **Communicate openly:** Share relevant information about your diabetes history, symptoms, medications, allergies, and lifestyle habits with healthcare providers. Describe any changes in health status, emotional well-being, or concerns affecting diabetes management.
- **Discuss goals:** Collaborate with healthcare providers to establish realistic goals for blood sugar control, weight management, physical activity, nutrition, and overall well-being. Set priorities based on personal preferences, values, and health priorities.
- **Participate in decision-making:** Engage in shared decision-making with healthcare providers to explore treatment options, weigh potential benefits and risks, and make informed decisions that align with your preferences and health goals.
- **Seek clarification:** Request explanations or clarifications about medical terminology, test results, treatment protocols, or recommended lifestyle changes. Understand the rationale behind healthcare recommendations and ask for additional resources or educational materials.
- **Advocate for access:** Advocate for timely access to diabetes medications, supplies, laboratory tests, referrals to specialists, and preventive care services.

Address barriers to healthcare access, such as insurance coverage issues or prior authorization requirements.

- **Follow up:** Follow through with recommended follow-up appointments, laboratory tests, screenings, and diabetes education sessions. Monitor progress toward achieving diabetes management goals and address any challenges or setbacks promptly.
- **Document information:** Keep a record of healthcare appointments, test results, medication changes, and discussions with healthcare providers. Maintain a personal health journal or use digital tools to track diabetes-related data, symptoms, and self-care activities. By actively advocating for themselves in healthcare settings, individuals with diabetes can foster collaborative relationships with healthcare providers, receive personalized care, access essential resources, and achieve better diabetes management outcomes.

39. What is the role of exercise in diabetes management?

Exercise plays a critical role in diabetes management by offering multiple benefits:

- **Blood sugar control:** Physical activity helps to lower blood glucose levels by increasing insulin sensitivity, allowing cells to use glucose more effectively.
- **Cardiovascular health:** Regular exercise improves cardiovascular fitness, lowers blood pressure, and reduces the risk of heart disease, which is common among people with diabetes.
- **Weight management:** Physical activity contributes to weight loss or maintenance, which is important for improving insulin sensitivity and reducing the risk of complications associated with obesity.

40. How can people with diabetes stay active during bad weather?

During inclement weather, individuals with diabetes can maintain physical activity indoors with:

- **Home workouts:** Engage in activities such as bodyweight exercises, yoga, or using exercise equipment like treadmills or stationary bikes.

- **Gym exercises:** Utilize indoor gym facilities for strength training, aerobic exercises, or fitness classes tailored to personal fitness levels.
- **Active hobbies:** Pursue indoor activities like dancing, swimming in indoor pools, or participating in recreational sports leagues to stay active and enjoy physical movement.

41. What are some diabetes-friendly recipes?

Diabetes-friendly recipes emphasize balanced nutrients and low sugar content, such as:

- **Grilled chicken with vegetables:** A lean protein source paired with fiber-rich vegetables like broccoli and carrots.
- **Quinoa salad:** Incorporates whole grains, vegetables, and a light vinaigrette dressing.
- **Baked fish with steamed greens:** Provides omega-3 fatty acids and vitamins, served with leafy greens or green beans.

42. How can people with diabetes manage their condition during stressful times?

Managing diabetes during stressful periods involves:

- **Stress management techniques:** Practicing relaxation exercises, meditation, or deep breathing to reduce stress hormone levels and stabilize blood sugar.
- **Maintaining routines:** Establishing consistent meal times, medication schedules, and sleep patterns to maintain stable blood glucose levels.
- **Seeking support:** Connecting with healthcare providers, diabetes educators, or support groups to discuss stress-related challenges and receive guidance.

43. What are the benefits of regular checkups for people with diabetes?

Regular checkups offer several advantages:

- **Health monitoring:** Healthcare providers assess blood sugar control, blood pressure, cholesterol levels, and screen for diabetes-related complications like eye disease or kidney problems.

- **Complication management:** Early detection of complications allows for timely intervention and treatment adjustments to prevent or minimize long-term health issues.
- **Treatment optimization:** Healthcare providers review diabetes management strategies, adjust medication regimens, and provide education on lifestyle modifications based on individual health needs.

44. How can people with diabetes find reliable information?

Reliable information sources include:

- **Healthcare providers:** Consult primary care physicians, endocrinologists, dietitians, or certified diabetes educators for personalized guidance and recommendations.
- **Reputable websites:** Access information from trusted organizations such as the American Diabetes Association (ADA), Centers for Disease Control and Prevention (CDC), or Mayo Clinic, which offer evidence-based resources and educational materials.
- **Diabetes organizations:** Engage with organizations dedicated to diabetes advocacy and support, which provide reliable information, community forums, and updates on diabetes management best practices.

45. What are some myths about diabetes management?

Common myths include:

- **Sugar consumption:** The myth that people with diabetes cannot consume any sugar is inaccurate. Moderation and carbohydrate counting are key to managing blood sugar levels effectively.
- **Insulin therapy stigma:** Believing that needing insulin therapy indicates failure in managing diabetes overlooks the fact that insulin is a vital treatment for many individuals to achieve optimal blood sugar control.

46. How can people with diabetes manage their condition at work?

Managing diabetes at work involves:

- **Meal planning:** Prepare balanced meals and snacks that align with dietary needs and blood sugar management goals. Consider portion sizes and timing to maintain stable glucose levels throughout the workday.
- **Staying active:** Incorporate physical activity into daily routines, such as taking short walks during breaks or using stairs instead of elevators, to promote circulation and regulate blood sugar.
- **Communicating needs:** Inform employers or colleagues about diabetes management requirements, such as scheduled breaks for blood glucose monitoring or insulin administration, to ensure a supportive work environment.

47. What is the importance of hydration for people with diabetes?

Hydration is crucial for individuals with diabetes because:

- **Blood sugar control:** Staying hydrated helps to regulate blood glucose levels by supporting kidney function and reducing the risk of dehydration-related blood sugar spikes.
- **Overall health:** Proper hydration promotes cardiovascular health, aids digestion, and supports overall well-being, which is beneficial for managing diabetes-related complications.

48. How can people with diabetes build a strong healthcare team?

Building a comprehensive healthcare team involves:

- **Primary care physician:** Manages overall health, performs routine screenings, and coordinates diabetes care with specialists.
- **Endocrinologist:** Specializes in diabetes management, adjusts medication regimens, and monitors diabetes-related complications.

- **Registered dietitian:** Provides nutrition counseling, creates personalized meal plans, and educates on carbohydrate counting and healthy eating habits.
- **Diabetes educator:** Offers self-management education, teaches blood glucose monitoring techniques, and provides ongoing support for lifestyle modifications.
- **Other specialists:** Includes ophthalmologists for eye health, podiatrists for foot care, and mental health professionals for emotional well-being, addressing comprehensive diabetes care needs.

By leveraging these resources and strategies, individuals with diabetes can effectively manage their condition, enhance quality of life, and reduce the risk of complications associated with diabetes.

49. Can diabetes become prevalent in minority populations?

Yes, diabetes has the potential to be prevalent within minority populations. Many factors can contribute to this result: genetics, sedentary lifestyle, and poor nutrition, to mention a few.

- **Genetics:** Some minority groups have genetic factors that make them more likely to develop diabetes. For instance, some people can have the tendency to develop fat around the abdomen, which increases the chance of acquiring diabetes.
- **Lifestyle:** Leading a sedentary lifestyle is a common cause of developing diabetes.
- **Poor nutrition:** Obtaining essential nutrition is vital for the human body. Eating unhealthy foods can cause malfunction of beta cells and spike blood pressure, which can contribute to the development of diabetes.

Diabetes indeed has a higher prevalence in certain minority populations, resulting from a complex interplay between genetic predisposition, lifestyle, and nutritional factors. In genetics, it has been well established that certain minority groups carry genetic markers that increase susceptibility to Type 2 diabetes. For instance, some studies have found that people of Hispanic, African American, and Indigenous descent may have a predisposition to gain fat in the abdominal area, something that strongly correlates with insulin resistance. Insulin resistance is one of the major risk factors for the development of diabetes since it makes the body unable to utilize insulin properly and hence result in high blood sugar levels. Each of these genetic factors, added to the environmental pressures usually felt by minority communities, increases the risk of diabetes.

Lifestyle choices also play a huge role in diabetes rates among minority groups. Socioeconomic disparities may mean that minorities do not have the same access to safe outdoor areas, making physical activity difficult. A sedentary lifestyle—often an accompaniment to urban living situations where walkways, parks, or play areas are in short supply—also contributes to weight gain and insulin resistance. Further, many minority group members work long hours, leaving little chance for physical activity. Those lifestyle challenges are further compounded by limited health education or awareness of the importance of regular physical activity, which creates a cycle in which inactivity becomes normative and health risks multiply.

Nutrition is another very strong contributing factor to diabetes prevalence in minority populations. Often in minority communities "there exist "food deserts"; these are " areas in which finding affordable healthy foods is a challenge. Such settings are not quite ideal for healthy nutrition, and most turn to fast foods that are loaded with sugars and fats. Poor diet also causes strain on the pancreatic beta cells which secrete insulin. This eventually diminishes their functioning and exposes one to the risk of developing diabetes. A diet mainly consisting of processed foods loaded with added salt also increases blood pressure levels and cholesterol levels, both of which exacerbate diabetes and are generally unhealthy. Together, these create a situation where diabetes prevalence is more than just a medical problem; it projects broader social and economic issues that concern minority health.

50. How does diabetes impact bone health?

Diabetes—particularly Type 1 and poorly managed Type 2—can have a weakening effect on bone density, which increases the risk of fractures. Glucose can damage the cells responsible for the formation of bone, and some medications used in treating diabetes may affect bone as well. Weight-bearing exercises, adequate intake of calcium and vitamin D, and regular checkins with a healthcare provider can help keep bones strong and prevent osteoporosis in diabetes.

Diabetes also impacts bones with various mechanisms, such as chronic inflammation and impaired calcium metabolism, which weakens the bone structure little by little. The high blood sugar levels in diabetes lead to oxidative stress and inflammation, which may slow down the body's process of renewing bone cells. In addition, insulin plays some role in bone formation, and in subjects with insulin resistance or low insulin production, there might be a reduction in bone mass and slowing of the healing process in cases of fractures. This is most prominent in patients with Type 1 diabetes since they are already disadvantaged by having no insulin and also autoimmune components, which may destroy bone health from young age onwards.

Some diabetes medications also impact bone density. For instance, thiazolidinediones—a class of drugs commonly used in Type 2 diabetes—were, by some studies, found to possibly lead to a reduction in bone mineral density, which would thus increase the likelihood of fractures. The lifestyle interventions, on the other hand—for example, regular weight-bearing exercises with or without resistance—are proven to enhance bone density and overall bone health. A diet with calcium and vitamin D, among other nutrients, contributes to the strength of bones; screening can also easily detect bone loss early in order to make interventions on time. Most benefited from this integrated approach are individuals with diabetes, wherein effective management of diabetes is coupled with proactive bone health strategies that would minimize the risk of fractures and osteoporosis.

51. How does diabetes impact our digestion process?

Diabetes can damage the vagus nerve, which helps control stomach muscles, leading to a condition called gastroparesis. This can cause symptoms like nausea, vomiting, bloating, and unpredictable blood sugar levels. Treatment may include dietary adjustments, medications to enhance gastric motility, and close monitoring of blood glucose. Small, frequent meals and avoiding high-fat or high-fiber foods can help manage symptoms.

Gastroparesis, also called "delayed gastric emptying," is a serious complication of diabetes. It develops when long-standing high levels of blood glucose cause damage to the vagus nerve-the very important nerve that controls contraction of stomach muscles. Once the vagus nerve is damaged, food moves slowly, or sometimes not at all, from the stomach into the small intestine. This delayed-development can produce heavy discomfort, influencing quality of life and making blood sugar management difficult to handle, since unpredictable emptying into the intestine begets spikes of glucose at ungodly hours. Gastroparesis is more common in persons with long-standing Type 1 or Type 2 diabetes, and its early diagnosis and management are important in order not to allow the symptomatology to worsen.

Key management strategies for gastroparesis include the following:

- Eating small amounts often
- Avoid high-fiber and high-fat foods
- Adequate hydration, fluid intake between meals
- Opting for softer, well-cooked foods
- Close monitoring of blood sugar

These dietary changes are important in that small, frequent meals reduce the bulk in the stomach and hence are easy to digest, thus minimizing the risk of bloating or nausea. Foods that are rich in fiber and fat are not easily digested; therefore, their avoidance may prevent them from staying in the stomach for a very short time.

Hydration and fluid intake between, rather than during, meals support easier digestion as well, since liquids can dilute the digestive elements and affect the speed at which the stomach will empty. Simpler meals of soft, well-cooked foods, such as steamed vegetables and grains, allow more comfortable and easily tolerable digestion, thereby reducing symptoms. Individuals with gastroparesis really need to be very particular with their blood sugar monitoring, as the erratic gastric emptying could give them some really crazy highs and lows. In some instances, physicians may prescribe medications that help improve gastric motility, which is very much associated with symptomatic relief. Symptoms are managed with medications and lifestyle modification; close follow-up with a healthcare provider is really important, where regular review and adjustment of the treatment stay crucial for symptom and glucose control.

52. What should people with diabetes know before consuming alcohol?

Alcohol has a very complicated effect on blood sugar: It can raise and lower glucose levels, therefore it's somewhat unpredictable to monitor in relation to diabetes management. Alcohol must first be processed within the liver, which is also a key organ responsible for the maintenance of blood sugar through the release of glucose into the blood. The liver, since it focuses primarily on breaking down the alcohol, temporarily ceases or reduces the releasing of glucose. This would drop the blood sugar level much further down, especially in those on insulin or on medications lowering glucose. This retarded glucose absorption may increase the risks of hypoglycemia in diabetic patients, especially those on insulin, occurring hours after consumption and sometimes even at night, hence complicating the management of their blood sugar.

Alcohol, therefore, needs to be handled extremely carefully by individuals with diabetes due to such risks. Another advantage to drinking at a meal is that absorption of both the alcohol and the carbohydrates is slowed, avoiding extreme peaks and valleys in blood glucose. Thus, fewer alcoholic beverages, or less often, lighten the liver's workload, freeing it up to do more work maintaining blood glucose.

This should be measured before and well many hours after the drink since hypoglycemia may appear so much later. It is most true when people are drunk on an empty stomach. If available, the added use of a continuous glucose monitor will also aid in the detection of any late fall of the blood sugar and serves as added protection.

53. Can the environment around us be the reason for our diabetes?

Environmental factors such as pollution, socioeconomic conditions, and lifestyle choices are increasingly recognized as significant contributors to the development of diabetes, especially Type 2 diabetes. A major concern is air pollution, which research indicates is linked to higher rates of insulin resistance, a crucial element in Type 2 diabetes. Pollutants like particulate matter (PM2.5) and nitrogen dioxide (NO_2) can lead to systemic inflammation and oxidative stress, both of which hinder insulin sensitivity and disrupt glucose metabolism. Regular exposure to these pollutants can make it difficult for individuals to use insulin effectively, resulting in elevated blood glucose levels. This risk is particularly heightened in urban areas with high pollution, impacting entire communities over time.

In addition to pollution, other environmental factors such as access to nutritious foods, safe spaces for exercise, and socioeconomic status also play a vital role in diabetes risk. For example, people in lower-income neighborhoods may find it challenging to access fresh, healthy foods and often reside in "food deserts," where fast-food outlets and convenience stores dominate. This situation promotes a diet rich in processed foods and sugary drinks, which can lead to obesity and insulin resistance. Furthermore, neighborhoods that lack parks or recreational areas tend to discourage physical activity, contributing to a more sedentary lifestyle that increases the risk of diabetes.

Chemical exposure is yet another environmental factor associated with diabetes. Endocrine-disrupting chemicals (EDCs), commonly found in plastics, pesticides, and personal care items, can interfere with the body's hormonal balance, including insulin regulation. Substances like bisphenol A (BPA) and phthalates may mimic or disrupt hormones, potentially resulting in insulin resistance. These chemicals are prevalent and often unavoidable, making it challenging for individuals to completely limit their exposure. Over time, the cumulative effects can be significant.

54. How does diabetes affect dental health and dental hygiene?

Primarily, diabetes can affect a person's dental health and hygiene due to the body's incapacity to control blood glucose levels. Several of these effects are outlined below, with evidence from reputable authorities to support these claims:

Periodontal Disease: Oral diseases caused by the infection and inflammation of the supportive gum and bone tissues which are referred to as periodontitis is believed to be diabetic. Focal excessive blood sugar results in diffuse excessive tissue glucose, leading to excessive saliva nutri-dense nourishment of gingivitis causing many periodontal diseases.

Dental Caries: Contrary to the notion that diabetes isn't associated with sugar consumption and caries, a few although heterogeneous empirical studies suggest that d iabetes, particularly where management is poor, may increase risk of dental caries. Such could be due to many factors like the side effects of the medication used (like dry mouth associated with some diabetes medications), change in saliva composition, and possibly, the impact of glucose per se on tooth enamel.

Dry Mouth (Xerostomia): A wide range of diabetes medications also elicits dry mouth as a side effect. Saliva is important for teeth protection by neutralizing acids, bond strengthening, and food digestion. Since they produce less saliva, they are more susceptible to tooth and gum decay as well as infections in the mouth.

Inhibition of Healing of the Body and Possible Enhancement of Infections: When sugar levels in blood are high, the body's ability to heal various wounds among them being the oral wounds is affected. Higher complications may be associated with Dental surgery, extractions, and even placing the dental implants in over uncontrolled diabetes patients. Furthermore, the propensity to have infections adds a barrier into the management of oral health.

Gingivitis, and its Rapid Progression to Periodontitis: Diabetic patients tend to hasten the switch from gingivitis, which is the inflammation of the gums, to periodontitis which is the infection of the gums and the supporting bone. The rapid change emphasizes the frequent dental visits required for diabetic patients.

55. How does circadian rhythm disruption and irregular sleep contribute to diabetes?

Our circadian rhythms, or our body clocks, play a part in how we metabolize sugar and how effectively we synthesize insulin throughout the day. Our bodies efficiently handle food intake during the times when people stand to be active. But when people disrupt this cycle with irregular sleeping patterns or other forms of stress, their metabolic competence plummets. Studies have shown that night shift workers are 40 percent more likely to develop type 2 diabetes relative to their day working counterparts.

The role of sleep in the development of certain diseases, in this case diabetes, appears to be mediated by a hormone called melatonin, otherwise known as the 'sleep hormone'. Mongeflow and his team have discovered that beta cells in the pancreas that produce insulin also seemed to have receptors for melatonin. Insulin levels in the body tend to be low at night when melatonin levels are high; because there shouldn't be a great need for food at these times.

With regard to the treatment of diabetes, time is everything, not only with drugs but with food and physical activity as well. A growing body of healthcare literature advises that diabetes patients' chronobiology should also be catered for as well; this gives rise to "chronotherapy" which involves treating patients according to their circadian cycles.

Conclusion

In conclusion, living with diabetes requires a proactive and integrated approach that spans various facets of daily life. By embracing education, regular monitoring, and personalized management plans, individuals can effectively manage blood sugar levels and reduce the risk of complications. Additionally, maintaining a balanced diet, engaging in regular physical activity, managing stress, and seeking support from healthcare providers and peers are pivotal in achieving optimal health outcomes. As research continues to evolve and technological advancements offer new tools, the landscape of diabetes management continues to improve, promising a brighter future for those living with the condition. Ultimately, by fostering resilience, determination, and a supportive network, individuals with diabetes can navigate their journey with confidence and lead fulfilling lives.

Bibliography

1. American Diabetes Association. (n.d.). *It's a great time to get moving.* https://diabetes.org/health-wellness/fitness
2. Mayo Clinic. (2024, June 11). *Diabetes diet: Create your healthy-eating plan.* https://www.mayoclinic.org/diseases-conditions/diabetes/in-depth/diabetes-diet/art-20044295
3. Bauer, G. (2024, October 4). *Stress SOS.* Diabetes Canada. https://www.diabetes.ca/about-diabetes/impact-stories/stress-sos
4. American Diabetes Association. (n.d.). *Checking your blood glucose.* https://www.diabetes.org/diabetes/medication-management/blood-glucose-testing-and-control/checking-your-blood-glucose
5. American Diabetes Association. (n.d.). *Diabetes basics.* https://www.diabetes.org/diabetes
6. Diabetes Canada (2023, December 8). *Debunking diabetes myths.* https://www.diabetes.ca/about-diabetes-(3)/impact-stories/debunking-diabetes-myths
7. Canadian Centre for Occupational Health and Safety. (2025). *Diabetes in the workplace.* Government of Canada. https://www.ccohs.ca/oshanswers/diseases/diabetes.html
8. Mayo Clinic. (2022). *Water: How much should you drink every day?* https://www.mayoclinic.org/healthy-lifestyle/nutrition-and-healthy-eating/in-depth/water/art-20044256
9. American Diabetes Association. (n.d.). *Your health care team.* https://diabetes.org/health-wellness/diabetes-and-your-health/your-healthcare-team
10. Murray, C. E., & Coleman, C. M. (2019). Impact of d iabetes m ellitus on b one h ealth. *International J ournal of M olecular S ciences, 20*(19), 4873. https://doi.org/10.3390/ijms20194873
11. Preshaw, P. M., Alba, A. L., Herrera, D., Jepsen, S., Konstantinidis, A., Makrilakis, K., & Taylor, R. (2012). Periodontal disease and diabetes: A two-way relationship. *Diabetologia, 55*(1), 21–31. https://doi.org/10.1007/s00125-011-2342-y
12. Priyanka, K., Sudhir, K. M., Reddy, C. V. K., Kumar, R. K., & Srinivas, M. (2016). Diabetes and oral health problems: An overview. *Journal of Clinical and Diagnostic Research, 10*(7), ZE11–ZE14. https://doi.org/10.7860/JCDR/2016/18659.8164
13. Mayo Clinic Staff. (2023, August 26). *Dry mouth.* Mayo Clinic. https://www.mayoclinic.org/diseases-conditions/dry-mouth/symptoms-causes/syc-20356048
14. Spampinato, C., Tashjian, S. K., Deng, C. Z. C., D'Souza, A. M. V. D. S., & Veves, A. (2021). Diabetes and wound healing: Pathophysiology, clinical challenges, and future directions. *International Journal of Molecular Sciences, 22*(19), 10435. https://doi.org/10.3390/ijms221910435

Diabetes and Nutrition

CHAPTER 3

Introduction

Navigating nutrition is pivotal for individuals managing diabetes, as dietary choices profoundly impact blood sugar control, overall health, and quality of life. This chapter delves into the intricacies of nutrition management tailored specifically for people with diabetes, offering comprehensive insights into optimizing dietary habits to promote well-being and mitigate health risks.

Understanding how different nutrients affect blood sugar levels is fundamental in diabetes management. Carbohydrates, for instance, directly influence glucose levels postconsumption, making carbohydrate intake management a cornerstone of daily dietary planning. This chapter explores strategies for choosing the right types and amounts of carbohydrates, emphasizing whole grains, fruits, vegetables, legumes, and dairy products as optimal choices that promote stable blood sugar levels.

Protein intake is another critical component discussed here, as it not only supports muscle health but also helps in stabilizing blood sugar and promoting satiety. Healthy protein sources such as lean meats, fish, eggs, beans, nuts, and seeds are highlighted for their role in balanced nutrition and overall wellness.

Additionally, the chapter delves into the role of fats in diabetes management, stressing the importance of consuming healthy fats like those found in avocados, nuts, olive oil, and fatty fish in moderation. These fats contribute to cardiovascular health and overall well-being, complementing a diet designed to support diabetes management.

Fiber's significant impact on blood sugar control and digestion is also explored, with an emphasis on incorporating fiber-rich foods such as fruits, vegetables, whole grains, legumes, and nuts into daily meals. Practical guidance on portion control, glycemic index considerations, and the benefits of meal planning further empower individuals to make informed dietary choices that align with their health goals.

Throughout this chapter, readers will find actionable tips, meal ideas, and strategies to enhance their dietary habits, ensuring they can navigate social events, dining out, and everyday challenges while effectively managing their diabetes through nutrition. By adopting these principles and strategies, individuals can cultivate a balanced, sustainable approach to eating that supports long-term health and well-being.

This introduction sets the stage by highlighting key topics related to nutrition management in diabetes, emphasizing practical advice and strategies that will be explored further in the chapter.

1. How does carbohydrate intake affect blood sugar levels?

Carbohydrates are broken down into glucose, which raises blood sugar levels, so managing carb intake is crucial.

2. What are good carbohydrate choices for people with diabetes?

Good choices include whole grains, fruits, vegetables, legumes, and low-fat dairy products.

3. How does protein intake impact diabetes management?

Protein helps to stabilize blood sugar levels and promote satiety, making it an important part of a balanced diet.

4. What are healthy sources of protein for people with diabetes?

Healthy sources include lean meats, fish, eggs, beans, nuts, and seeds.

5. How does fat intake influence diabetes management?

Healthy fats can help to manage cholesterol levels and support overall health, but they should be consumed in moderation.

6. What are healthy fat sources for people with diabetes?

Healthy fat sources include avocados, nuts, seeds, olive oil, and fatty fish.

7. How can fiber benefit people with diabetes?

Fiber slows down the absorption of sugar, helping to control blood sugar levels and improve digestion.

8. What are good sources of fiber for people with diabetes?

Good sources include fruits, vegetables, whole grains, legumes, and nuts.

9. How important is portion control for diabetes management?

Portion control helps to manage blood sugar levels and maintain a healthy weight.

10. How can people with diabetes practice portion control?

By using smaller plates, measuring food, and being mindful of serving sizes.

11. What is the role of glycemic index in diabetes management?

The glycemic index ranks foods based on how quickly they raise blood sugar levels, helping people choose foods that have a lower impact on blood sugar.

12. What are some low-glycemic foods?

Low-glycemic foods include most vegetables, some fruits, legumes, and whole grains.

13. How can meal planning help in managing diabetes?

Meal planning ensures balanced nutrition, proper portion sizes, and consistent carbohydrate intake.

14. What is carb counting and why is it important?

Carb counting involves tracking carbohydrate intake to manage blood sugar levels effectively.

15. How can people with diabetes enjoy eating out?

By choosing balanced meals, controlling portion sizes, and asking for nutritional information when available.

16. What are some healthy snack options for people with diabetes?

Healthy snacks include nuts, seeds, yogurt, fruits, and vegetables.

17. How does sugar impact diabetes management?

Sugar can cause rapid spikes in blood sugar levels, so it should be consumed in moderation.

18. What are some alternatives for sugar for people with diabetes?

Alternatives include natural sweeteners like stevia, erythritol, and monk fruit.

19. How can people with diabetes manage cravings for sweets?

By choosing healthier alternatives, eating balanced meals, and practicing mindful eating.

20. What is the importance of hydration in diabetes management?

Staying hydrated helps to control blood sugar levels and maintain overall health.

21. What are some diabetes-friendly beverages?

Diabetes-friendly beverages include water, unsweetened tea, and low-calorie drinks.

22. How can alcohol consumption affect diabetes management?

Alcohol can cause fluctuations in blood sugar levels and interact with diabetes medications.

23. What should people with diabetes know about dietary supplements?

Supplements should be taken under medical supervision, as some can interact with diabetes medications.

24. How can people with diabetes navigate grocery shopping?

By reading labels, choosing whole foods, and avoiding processed foods with added sugars.

25. What types of recipes are diabetes-friendly?

Recipes that include balanced nutrients, low sugar content, and healthy ingredients.

26. How can people with diabetes manage their diet during holidays and special occasions?

By planning ahead, making healthy choices, and monitoring blood sugar levels.

27. What is the importance of breakfast for people with diabetes?

Eating a balanced breakfast helps to stabilize blood sugar levels throughout the day.

28. What are some healthy breakfast options for people with diabetes?

Healthy options include oatmeal, eggs, whole-grain toast, and Greek yogurt.

29. How can people with diabetes manage their diet while traveling?

By packing healthy snacks, choosing balanced meals, and staying hydrated.

30. What are some tips for cooking diabetes-friendly meals at home?

Use fresh ingredients, control portion sizes, and avoid added sugars and unhealthy fats.

31. How can people with diabetes reduce their sodium intake?

By choosing fresh foods, reading labels, and avoiding processed and packaged foods.

32. What is the role of vitamins and minerals in diabetes management?

Vitamins and minerals support overall health and can help to manage blood sugar levels.

33. How can people with diabetes ensure they get enough vitamins and minerals?

By eating a varied diet that includes fruits, vegetables, whole grains, and lean proteins.

34. What is the impact of caffeine on diabetes management?

Caffeine can affect blood sugar levels, so it should be consumed in moderation.

35. How can people with diabetes enjoy desserts?

By choosing healthier alternatives, controlling portions, and using natural sweeteners.

36. What are some healthy lunch ideas for people with diabetes?

Healthy lunch ideas include salads, whole grain sandwiches, and vegetable stir-fries.

37. How can people with diabetes handle eating at social events?

By planning ahead, making healthy choices, and monitoring blood sugar levels.

38. What are some diabetes-friendly cooking methods?

Healthy cooking methods include baking, grilling, steaming, and sautéing.

39. What are some tips for heart-healthy cooking with diabetes?

By choosing lean proteins, using healthy oils, and avoiding fried and processed foods.

40. How can people with diabetes manage their diet on a budget?

By planning meals, buying in bulk, and choosing affordable, nutritious foods.

41. What is the role of probiotics in diabetes management?

Probiotics can support gut health and may help to improve blood sugar control.

42. How can people with diabetes incorporate probiotics into their diet?

By eating foods like yogurt, kefir, sauerkraut, and other fermented foods.

43. What are some healthy dinner ideas for people with diabetes?

Healthy dinner ideas include grilled chicken with vegetables, salmon with quinoa, and vegetable stir-fry.

44. How can people with diabetes avoid added sugars?

By reading labels, choosing whole foods, and avoiding processed foods and sugary drinks.

45. What are some tips for eating mindfully with diabetes?

Pay attention to hunger and fullness cues, eat slowly, and savor each bite.

46. How can people with diabetes balance their meals?

By including a mix of carbohydrates, proteins, and healthy fats in each meal.

47. What are some diabetes-friendly meal prep tips?

Plan meals ahead, cook in batches, and store meals in portion-controlled containers.

48. How can people with diabetes handle food cravings?

By choosing healthier alternatives, staying hydrated, and practicing mindful eating.

49. What is the importance of a varied diet for people with diabetes?

A varied diet ensures a wide range of nutrients and helps to prevent diet monotony.

50. How can people with diabetes stay motivated to eat healthily?

By setting realistic goals, tracking progress, and seeking support from healthcare providers and support groups.

51. How can people with diabetes protect their vision and prevent retinopathy?

Diabetic retinopathy, damage to the blood vessels in the retina, is one of the leading causes of vision loss among people with diabetes. Preventive measures and early detection can significantly reduce the risk:

- **Control blood sugar and blood pressure:** Keeping blood glucose and blood pressure levels within recommended ranges reduces stress on blood vessels, protecting retinal health.
- **Regular eye exams:** An annual comprehensive eye exam with pupil dilation allows eye specialists to detect early signs of retinopathy and other eye issues.
- **Monitor cholesterol levels:** Elevated cholesterol contributes to vessel damage. A healthy diet, regular exercise, and medications (if prescribed) can help maintain cholesterol levels.
- **Avoid smoking:** Smoking increases the risk of blood vessel damage, exacerbating retinopathy risk. Quitting smoking benefits both retinal and overall vascular health.
- **Laser therapy and injections:** For individuals with early signs of retinopathy, laser treatment can seal leaking blood vessels. Anti-VEGF injections reduce abnormal vessel growth in more advanced cases.

52. What lifestyle changes are most effective in preventing diabetic nephropathy?

Diabetic nephropathy, or kidney damage caused by diabetes, is a significant risk for people with both Type 1 and Type 2 diabetes. Adopting healthy lifestyle practices can help preserve kidney function:

- **Control blood glucose and blood pressure:** These are critical factors in preventing nephropathy, as high blood sugar and hypertension strain kidney function.

- **Limit protein intake:** While protein is essential, excessive intake can stress kidneys. A dietitian can help individuals determine appropriate protein levels.
- **Stay hydrated:** Proper hydration helps kidneys filter waste and prevents kidney stones.
- **Avoid excessive salt:** Reducing salt intake lowers blood pressure, protecting kidney health. Processed foods and restaurant meals are often high in sodium, so opting for fresh, homemade meals is beneficial.
- **Avoid nephrotoxic substances:** Certain medications, especially nonsteroidal anti-inflammatory drugs (NSAIDs) like ibuprofen, can harm kidneys. Consult healthcare providers before taking new medications.
- **Regular kidney function tests:** Routine tests, such as urine albumin-to-creatinine ratio (ACR) and estimated glomerular filtration rate (eGFR), detect early kidney damage and enable timely interventions.

Conclusion

In conclusion, managing diet plays a pivotal role in the effective management of diabetes, offering individuals the ability to regulate blood sugar levels, maintain a healthy weight, and enhance overall well-being. Throughout this exploration of dietary considerations for diabetes management, several key principles and strategies have emerged.

Understanding how different macronutrients impact blood sugar levels is fundamental. Carbohydrates, proteins, and fats each play unique roles in regulating metabolism and should be consumed in balanced proportions. Choosing complex carbohydrates like whole grains, fruits, and vegetables helps to stabilize blood sugar levels due to their slower digestion and absorption rates. Incorporating lean proteins such as fish, poultry, beans, and nuts supports muscle health and provides a sustained source of energy. Meanwhile, consuming healthy fats found in avocados, olive oil, and fatty fish contributes to heart health and satiety, although moderation is crucial to avoid excess calorie intake.

Fiber-rich foods also deserve emphasis for their ability to improve glycemic control and promote digestive health. Foods like legumes, nuts, whole grains, and vegetables not only help regulate blood sugar but also contribute to feelings of fullness and aid in weight management.

Portion control and mindful eating are essential practices for individuals with diabetes, helping to prevent overeating and maintain consistent blood sugar levels throughout the day. Monitoring carbohydrate intake through techniques like carb counting and understanding the glycemic index of foods assists in making informed dietary choices that support stable glucose levels.

Moreover, navigating social situations, dining out, and managing cravings for sweets or indulgent foods require thoughtful planning and adherence to dietary principles. By incorporating these strategies into daily routines and adopting a flexible approach to food choices, individuals with diabetes can enjoy a varied and satisfying diet while effectively managing their condition.

Overall, a well-balanced diet tailored to individual needs, combined with regular physical activity and ongoing support from healthcare providers, forms the cornerstone of successful diabetes management. By embracing these principles, individuals with diabetes can achieve optimal health outcomes, enhance their quality of life, and reduce the risk of complications associated with diabetes.

This conclusion encapsulates the importance of dietary management in diabetes, highlights key strategies for maintaining blood sugar control, and reinforces the role of a balanced diet in promoting overall health and well-being.

Bibliography

1. American Heart Association. (2024, April 2). *Cholesterol and diabetes.* *https://www.heart.org/en/health-topics/diabetes/diabetes-complications-and-risks/cholesterol-abnormalities--diabetes*
2. American Diabetes Association. (n.d.). *Eye complications.* *https://diabetes.org/about-diabetes/complications/eye-complications*
3. American Diabetes Association. (n.d.). *Foot care tips.* *https://diabetes.org/health-wellness/diabetes-and-your-feet/foot-care-tips#:~:text=Wash%20your%20feet%20well%20every,which%20can%20lead%20to%20infections.*
4. American Diabetes Association. (n.d.). *Diabetes technology guide.* *https://diabetes.org/living-with-diabetes/treatment-care/diabetes-technology-guide*
5. Government of Canada. (2024, January 12). *Tobacco and type 2 diabetes.* *https://www.canada.ca/en/health-canada/services/health-concerns/tobacco/legislation/tobacco-product-labelling/tobacco-type-2-diabetes.html*
6. American Veterinary Medical Association. (n.d.). *Diabetes in pets.* *https://www.avma.org/resources-tools/pet-owners/petcare/diabetes-pets*
7. Centers for Disease Control and Prevention. (2024, May 15). *About type 1 diabetes.* *https://www.cdc.gov/diabetes/about/about-type-1-diabetes.html*
8. Mayo Clinic. (2020, September 15). *Foot care with diabetes.* *https://connect.mayoclinic.org/blog/take-charge-healthy-aging/newsfeed-post/foot-care-with-diabetes/*
9. Mayo Clinic. (2025). *Gestational diabetes.* *https://www.mayoclinic.org/diseases-conditions/gestational-diabetes/diagnosis-treatment/drc-20355345*
10. Grossman, L. D., Roscoe, R., & Shack, A. R. (2018). Complementary and alternative medicine for diabetes. *Canadian Journal of Diabetes, 42*(Suppl 1), S154–S161. https://doi.org/10.1016/j.jcjd.2017.10.023
11. National Eye Institute. (2024). *Diabetic retinopathy.* *https://www.nei.nih.gov/learn-about-eye-health/eye-conditions-and-diseases/diabetic-retinopathy#:~:text=Diabetic%20retinopathy%20is%20caused%20by,vessels%20all%20over%20the%20body.*
12. National Institute of Diabetes and Digestive and Kidney Diseases. (2024). *Diabetes genetics & genomics.* *https://www.niddk.nih.gov/research-funding/research-programs/diabetes-genetics-genomics*
13. National Institute of Diabetes and Digestive and Kidney Diseases. (2021). *Diabetes, heart disease, and stroke.* *https://www.niddk.nih.gov/health-information/diabetes/overview/preventing-problems/heart-disease-stroke#:~:text=Adults%20with%20diabetes%20are%20nearly,stroke%20as%20adults%20without%20diabetes.&text=The%20good%20news%20is%20that,having%20heart%20disease%20or%20stroke*
14. National Institute of Diabetes and Digestive and Kidney Diseases. (2018, February). *What is diabetic neuropathy?* *https://www.niddk.nih.gov/health-information/diabetes/overview/preventing-problems/nerve-damage-diabetic-neuropathies/what-is-diabetic-neuropathy*

CHAPTER 4

Physical Activity and Diabetes

Introduction

Physical activity is a cornerstone of diabetes management, offering profound benefits that extend beyond blood sugar control to encompass overall health and well-being. In this chapter, we delve into the pivotal role of exercise in the lives of individuals living with diabetes, exploring its diverse forms, benefits, and practical considerations for integrating physical activity into daily routines.

For people with diabetes, regular physical activity plays a crucial role in lowering blood sugar levels, enhancing insulin sensitivity, and managing body weight. Aerobic exercises such as walking, cycling, swimming, and dancing help to improve cardiovascular health and can be easily tailored to individual fitness levels and preferences. Meanwhile, strength training exercises like weightlifting and resistance band workouts contribute to muscle strength and aid in better glucose metabolism. Additionally, flexibility exercises such as stretching and yoga improve joint mobility, reduce the risk of injuries, and promote overall flexibility.

Understanding how different types of exercises impact insulin sensitivity and blood sugar levels is essential for optimizing diabetes management. Equally important are strategies for managing blood sugar before, during, and after exercise, ensuring that physical activity remains safe and effective. Precautions such as monitoring blood sugar levels, staying hydrated, and having fast-acting sugar sources on hand are vital components of a successful exercise routine.

Moreover, this chapter explores practical tips for starting and maintaining an exercise regimen, adapting workouts to varying circumstances such as extreme weather or travel, and integrating physical activity into daily life and special occasions. Guidance on setting realistic goals, staying motivated, and leveraging support networks underscores the holistic approach to managing diabetes through physical activity.

As we navigate through the myriad benefits and considerations of exercise for diabetes management, it becomes evident that physical activity is not just a regimen but a lifestyle choice that empowers individuals to take control of their health and thrive. By embracing the principles outlined here, individuals with diabetes can embark on a journey of improved fitness, enhanced well-being, and sustained vitality.

This introduction sets the stage for exploring the importance of physical activity in diabetes management, emphasizing its benefits, various forms, and practical implementation strategies.

1. How does physical activity benefit people with diabetes?

Physical activity helps to lower blood sugar levels, improve insulin sensitivity, and maintain a healthy weight.

2. What types of exercise are recommended for people with diabetes?

Recommended exercises include aerobic activities, strength training, and flexibility exercises.

Figure 4.1
Yoga.

3. How often should people with diabetes exercise?

People with diabetes should aim for at least 150 minutes of moderate-intensity aerobic activity per week, plus strength training on two non consecutive days.

4. What precautions should people with diabetes take before starting an exercise program?

Consult a healthcare provider, check blood sugar levels, and prepare for managing low blood sugar.

5. How can people with diabetes monitor their blood sugar during exercise?

Check blood sugar levels before, during, and after exercise to understand its effects.

6. What should people with diabetes do if their blood sugar is too high or low before exercising?

Avoid exercise if blood sugar is above 250 mg/dL or below 70 mg/dL, and take corrective action.

7 . How can people with diabetes safely incorporate high-intensity interval training (HIIT)?

Begin with shorter, low-intensity intervals to build endurance, monitor blood sugar levels before, during, and after exercise. Consider consulting a healthcare provider to tailor the routine to individual health needs and to ensure safety.

8. What are the signs of low blood sugar during exercise?

Signs include shakiness, sweating, dizziness, confusion, and hunger.

9. How can people with diabetes safely start a new exercise routine?

Start slowly, increase intensity gradually, and consult a healthcare provider for guidance.

10. What role does strength training play in diabetes management?

Strength training builds muscle, improves insulin sensitivity, and supports weight management.

11. How can people with diabetes incorporate flexibility exercises into their routine?

Include yoga, stretching, or Pilates to improve flexibility, reduce stress, and prevent injuries.

12. What are some safe aerobic exercises for people with diabetes?

Walking, cycling, swimming, and dancing.

13. How can people with diabetes stay motivated to exercise regularly?

Set achievable goals, find enjoyable activities, and involve friends or fitness groups for support.

14. How can people with diabetes adjust their exercise routine if they experience joint pain?

Choose low-impact activities like swimming or cycling and consult a healthcare provider for advice.

15. What is the importance of warming up and cooling down in an exercise routine?

Warming up prepares the body for exercise, and cooling down helps lower the heart rate and prevent injuries.

16. How can people with diabetes ensure they stay hydrated during exercise?

Drink water before, during, and after exercise, especially in hot or humid conditions.

17. How does exercise affect insulin requirements for people with diabetes?

Exercise can lower insulin needs, requiring monitoring and dose adjustments with a healthcare provider.

18. What should people with diabetes consider when exercising in extreme temperatures?

Dress appropriately, stay hydrated, and avoid exercising in very hot or cold conditions.

19. How can people with diabetes manage exercise when feeling unwell?

Reduce intensity, rest, and consult a healthcare provider if necessary.

20. What are some ways to track progress in an exercise program?

Use a fitness tracker, maintain an exercise journal, or set measurable goals.

21. How can people with diabetes incorporate exercise into a busy schedule?

Break exercise into shorter sessions, use active transportation, or integrate physical activity into daily tasks.

22. What is the role of interval training in diabetes management?

Interval training improves cardiovascular fitness and blood sugar control in less time.

23. How can people with diabetes prevent foot problems during exercise?

Wear proper footwear, inspect feet for injuries, and maintain foot hygiene.

24. How can people with diabetes benefit from group fitness classes?

Group classes provide motivation, social interaction, and structured workouts.

25. What are some exercises that can be done at home for people with diabetes?

Bodyweight exercises, resistance band workouts, and online fitness classes.

26. How can high-intensity interval training (HIIT) help people with diabetes?

HIIT improves cardiovascular health, insulin sensitivity, and burns calories efficiently.

27. How can exercise help with stress management for people with diabetes?

Exercise helps with stress management by releasing endorphins, which improve mood and promote feelings of well-being. It also reduces stress hormones like cortisol, enhances relaxation, and can improve sleep, all of which contribute to better emotional balance and overall diabetes management.

28. How can people with diabetes balance exercise with other management strategies?

Coordinate exercise with meal planning, medication timing, and blood sugar monitoring.

29. What are the benefits of outdoor exercise for people with diabetes?

Outdoor activities provide fresh air, sunlight, and varied environments for motivation.

30. How can people with diabetes use technology to enhance their exercise routine?

You can use fitness apps, wearable trackers, and online resources to track progress and plan workouts.

31. How can people with diabetes handle setbacks in their exercise routine?

Practice self-compassion, adjust goals, and seek support from healthcare providers or peers.

32. What is the importance of rest days in an exercise program?

Rest days allow muscles to recover, prevent injuries, and maintain long-term motivation.

33. How can people with diabetes involve family and friends in their exercise routine?

Exercise together, participate in active outings, and support each other's goals.

34. How can people with diabetes find the right exercise intensity?

Use the talk test—moderate intensity allows speaking with slight breathlessness.

35. What are the benefits of swimming for people with diabetes?

Swimming improves cardiovascular fitness, supports joint health, and reduces impact on joints.

36. How can people with diabetes maintain an exercise routine during travel?

Plan ahead, use hotel gyms, and incorporate portable workouts into the schedule.

37. How can people with diabetes stay safe while exercising alone?

Inform someone of plans, carry identification and emergency glucose, and exercise in safe areas.

38. How can people with diabetes make exercise enjoyable?

Engage in fun activities, vary routines, and celebrate progress with rewards.

39. What are creative ways to stay active at work?

Take walking breaks, use a standing desk, or do simple exercises during meetings.

40. How can people with diabetes celebrate their fitness achievements?

Reward themselves with meaningful treats, share progress with friends, or set new goals.

Conclusion

In conclusion, physical activity stands as a cornerstone of effective diabetes management, offering a multitude of benefits that extend well beyond mere blood sugar control. Throughout this chapter, we have explored the diverse ways in which exercise contributes to the overall health and well-being of individuals living with diabetes.

One of the primary benefits of physical activity for people with diabetes is its ability to lower blood sugar levels and improve insulin sensitivity. Aerobic exercises such as walking, cycling, and swimming not only help in achieving better glucose control but also enhance cardiovascular health and contribute to weight management. Meanwhile, strength training exercises like weightlifting and resistance training build muscle strength, which is crucial for improving metabolism and overall physical function. Flexibility exercises such as yoga and stretching promote joint health, reduce the risk of injuries, and enhance mobility, making daily activities easier and more enjoyable.

Effective management of diabetes through physical activity requires careful planning and consideration of individual health needs. Strategies discussed in this chapter, such as monitoring blood sugar levels before, during, and after exercise, staying hydrated, and adjusting food intake and medication as necessary, are vital for maintaining safe and effective workouts.

Furthermore, integrating physical activity into daily routines and adapting to various circumstances like travel or weather conditions underscores the importance of consistency and flexibility in maintaining an active lifestyle. Setting realistic goals, finding activities that are enjoyable, and seeking support from healthcare providers, fitness professionals, or support networks are crucial elements in sustaining long-term exercise habits.

By embracing the principles and recommendations outlined in this chapter, individuals with diabetes can empower themselves to take control of their health, improve their quality of life, and reduce the risk of diabetes-related complications. Physical activity not only serves as a tool for managing diabetes but also fosters resilience, vitality, and overall well-being, demonstrating its indispensable role in the holistic approach to diabetes care.

This conclusion summarizes the key benefits of physical activity for diabetes management, highlights practical strategies for incorporating exercise into daily life, and emphasizes the empowering effect of an active lifestyle on overall health and well-being.

Bibliography

1. American Heart Association. (2024, April 2). *Cholesterol and diabetes.* https://www.heart.org/en/health-topics/diabetes/diabetes-complications-and-risks/cholesterol-abnormalities--diabetes
2. American Diabetes Association. (n.d.). *Eye complications.* https://diabetes.org/about-diabetes/complications/eye-complications
3. American Diabetes Association. (n.d.). *Foot care tips.* https://diabetes.org/health-wellness/diabetes-and-your-feet/foot-care-tips#:~:text=Wash%20your%20feet%20well%20every,which%20can%20lead%20to%20infections
4. American Diabetes Association. (n.d.). *Diabetes technology guide.* https://diabetes.org/living-with-diabetes/treatment-care/diabetes-technology-guide
5. Government of Canada. (2024, January 12). *Tobacco and type 2 diabetes.* https://www.canada.ca/en/health-canada/services/health-concerns/tobacco/legislation/tobacco-product-labelling/tobacco-type-2-diabetes.html
6. American Veterinary Medical Association. (n.d.). *Diabetes in pets.* https://www.avma.org/resources-tools/pet-owners/petcare/diabetes-pets
7. Centers for Disease Control and Prevention. (2024, May 15). *About type 1 diabetes.* https://www.cdc.gov/diabetes/about/about-type-1-diabetes.html
8. Mayo Clinic. (2020, September 15). *Foot care with diabetes.* https://connect.mayoclinic.org/blog/take-charge-healthy-aging/newsfeed-post/foot-care-with-diabetes/
9. Mayo Clinic. (2025, May 30). *Gestational diabetes.* https://www.mayoclinic.org/diseases-conditions/gestational-diabetes/diagnosis-treatment/drc-20355345
10. Grossman, L. D., Roscoe, R., & Shack, A. R. (2018, April). Complementary and alternative medicine for diabetes. *Canadian Journal of Diabetes, 42*(Suppl 1), S154–S161. https://doi.org/10.1016/j.jcjd.2017.10.023
11. National Eye Institute. (2024, December 10). *Diabetic retinopathy.* https://www.nei.nih.gov/learn-about-eye-health/eye-conditions-and-diseases/diabetic-retinopathy#:~:text=Diabetic%20retinopathy%20is%20caused%20by,vessels%20all%20over%20the%20body
12. National Institute of Diabetes and Digestive and Kidney Diseases. (2024, October). *Diabetes genetics & genomics.* https://www.niddk.nih.gov/research-funding/research-programs/diabetes-genetics-genomics
13. National Institute of Diabetes and Digestive and Kidney Diseases. (2021, April). *Diabetes, heart disease, & stroke.* https://www.niddk.nih.gov/health-information/diabetes/overview/preventing-problems/heart-disease-stroke#:~:text=Adults%20with%20diabetes%20are%20nearly,stroke%20as%20adults%20without%20diabetes.&text=The%20good%20news%20is%20that,having%20heart%20disease%20or%20stroke
14. National Institute of Diabetes and Digestive and Kidney Diseases. (2018). *What is diabetic neuropathy?* https://www.niddk.nih.gov/health-information/diabetes/overview/preventing-problems/nerve-damage-diabetic-neuropathies/what-is-diabetic-neuropathy

Medication Management in Diabetes

CHAPTER 5

Introduction

Chapter 5 delves into the intricacies of medication management for diabetes, a cornerstone in effectively controlling blood sugar levels and mitigating the risk of complications. Understanding the various types of medications available, their mechanisms of action, potential side effects, and best practices for usage is essential for individuals with diabetes and their healthcare providers. This chapter aims to equip readers with comprehensive knowledge and practical strategies to navigate the complexities of diabetes medication, ensuring optimal treatment outcomes and quality of life.

The management of diabetes through medications encompasses a diverse array of approaches, from insulin therapy to oral medications and non insulin injectables. Each type of medication plays a crucial role in addressing different aspects of glucose regulation, tailored to meet the unique needs of individuals based on their type of diabetes, overall health status, and lifestyle factors.

Effective medication management involves not only understanding how medications work but also adhering to prescribed schedules, monitoring blood sugar levels, and proactively addressing any potential side effects or interactions. Furthermore, this chapter explores the role of technology, such as insulin pumps and continuous glucose monitors (CGMs), in enhancing medication precision and improving diabetes management outcomes.

By delving into these topics, this chapter aims to empower individuals with diabetes and their caregivers to make informed decisions about medication choices, adhere to treatment plans effectively, and collaborate closely with healthcare providers to achieve optimal health outcomes. Whether managing medications during everyday life, travel, illness, or emergencies, understanding the principles outlined in this chapter can significantly enhance the management and quality of life for individuals living with diabetes.

1. What are the main types of diabetes medications?

Main types include insulin, oral hypoglycemics, and noninsulin injectables.

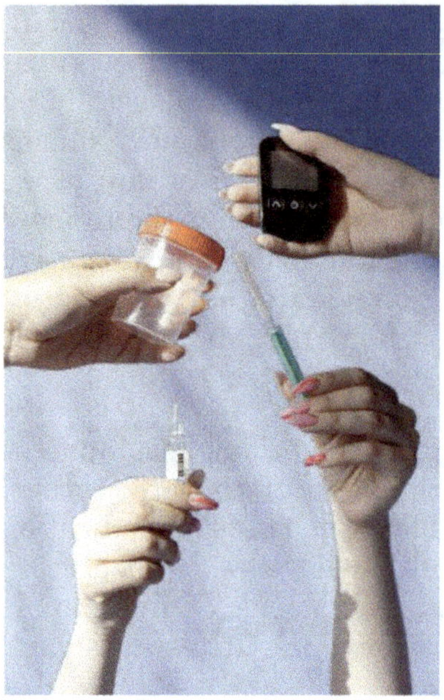

▲ Figure 5.1
Insulin monitoring.

2. How does insulin therapy work?

Insulin therapy involves administering insulin to help manage blood sugar levels in people whose bodies do not produce enough insulin.

3. What are the different types of insulin?

Types include rapid-acting, short-acting, intermediate-acting, and long-acting insulin.

4. How is insulin administered?

Insulin can be administered through injections, insulin pens, or insulin pumps.

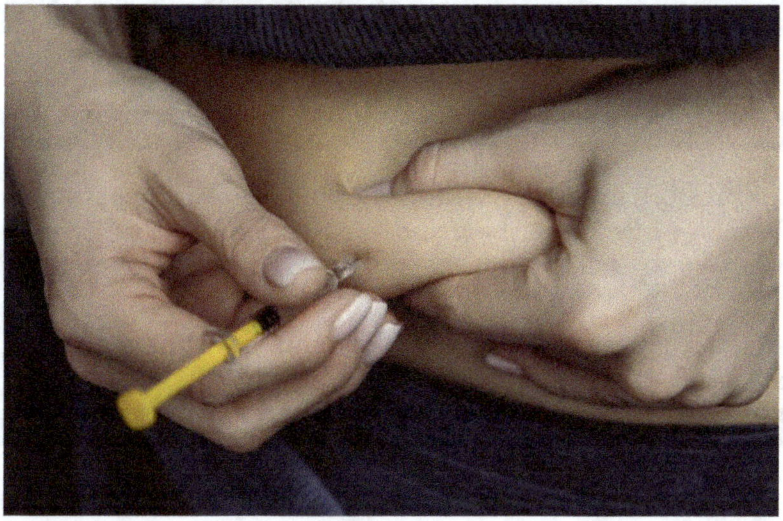

▲ Figure 5.2
Subcutaneous insulin injection.

5. What are some common oral medications for Type 2 diabetes?

Common oral medications include metformin, sulfonylureas, meglitinides, and DPP-4 inhibitors.

6. How does metformin work?

Metformin reduces glucose production in the liver and improves insulin sensitivity.

7. What are the side effects of metformin?

Side effects can include gastrointestinal issues like nausea, diarrhea, and abdominal discomfort.

8. How do sulfonylureas work?

Sulfonylureas stimulate the pancreas to release more insulin.

9. What are the side effects of sulfonylureas?

Side effects can include low blood sugar (hypoglycemia) and weight gain.

10. How do GLP-1 receptor agonists work?

GLP-1 receptor agonists increase insulin secretion, slow gastric emptying, and reduce appetite.

11. What are the side effects of GLP-1 receptor agonists?

Side effects can include nausea, vomiting, and diarrhea.

12. How do SGLT2 inhibitors work?

SGLT2 inhibitors prevent the kidneys from reabsorbing glucose, which is then excreted in the urine.

13. What are the side effects of SGLT2 inhibitors?

Side effects can include urinary tract infections, genital infections, and dehydration.

14. How do DPP-4 inhibitors work?

DPP-4 inhibitors help increase insulin release and decrease glucose production.

15. What are the side effects of DPP-4 inhibitors?

Side effects can include respiratory infections, headaches, and digestive issues.

16. How can people with diabetes manage medication schedules?

Use pill organizers, set reminders, and keep a medication log.

17. What should people with diabetes do if they miss a dose of medication?

Follow their healthcare provider's instructions, which may include taking the missed dose as soon as remembered or skipping it.

18. How can people with diabetes safely store their medications?

Store medications in a cool, dry place away from direct sunlight, and follow specific storage instructions for insulin.

19. What is the importance of medication adherence in diabetes management?

Adherence ensures effective blood sugar control and reduces the risk of complications.

20. How can people with diabetes reduce medication side effects?

Take medications with food, stay hydrated, and follow the prescribed dosage.

21. How can people with diabetes communicate effectively with their healthcare providers about medications?

Keep a list of all medications, report side effects, and ask questions about new prescriptions.

22. What is the role of a pharmacist in diabetes medication management?

Pharmacists can provide medication counseling, identify potential drug interactions, and offer tips for managing side effects.

23. How can people with diabetes handle medication adjustments?

Follow healthcare provider instructions and monitor blood sugar levels to assess the effectiveness of changes.

24. What is the impact of diet on diabetes medication effectiveness?

A balanced diet can enhance medication effectiveness and help stabilize blood sugar levels.

25. How can people with diabetes manage medications during travel?

Pack enough medication for the trip, keep medications in carry-on luggage, and carry a prescription list.

26. How do people with diabetes know if their medications are working?

Regular blood sugar monitoring and A1C tests help assess medication effectiveness.

27. How can people with diabetes safely use over-the-counter (OTC) medications?

Check for potential interactions with diabetes medications and consult with a healthcare provider.

28. What should people with diabetes know about herbal supplements?

Herbal supplements can interact with diabetes medications, so they should be used with caution and under medical supervision.

29. How can stress impact diabetes medication effectiveness?

Stress can affect blood sugar levels, potentially requiring medication adjustments.

30. What are some strategies for managing multiple medications?

Use a single pharmacy, maintain an updated medication list, and schedule regular medication reviews with a healthcare provider.

31. How can people with diabetes handle medication cost issues?

Look for generic options, apply for patient assistance programs, and discuss cost-saving options with a healthcare provider.

32. How do exercise and physical activity affect diabetes medications?

Exercise can improve insulin sensitivity and may require adjustments in medication dosages.

33. What are some common medication interactions with diabetes drugs?

Interactions can occur with blood pressure medications, cholesterol-lowering drugs, and certain antibiotics.

34. How can people with diabetes avoid medication errors?

Double-check labels, use a pill organizer, and confirm dosages with healthcare providers.

35. What is the importance of regular blood sugar monitoring in medication management?

It helps to assess how well medications are working and identify the need for adjustments.

36. How can people with diabetes manage their medications during illness?

Follow sick day plans provided by healthcare providers, monitor blood sugar more frequently, and adjust medications as needed.

37. What is the role of insulin pumps in diabetes management?

Insulin pumps deliver continuous insulin and can provide more precise blood sugar control.

38. How do continuous glucose monitors (CGMs) aid in medication management?

CGMs provide real-time blood sugar data, helping to make informed decisions about medication adjustments.

39. How can people with diabetes manage medication side effects like low blood sugar?

Carry a source of fast-acting sugar, monitor blood sugar levels regularly, and adjust medications under medical guidance.

40. What should people with diabetes do if they experience severe side effects from medication?

Seek immediate medical attention and inform their healthcare provider.

41. How can people with diabetes safely dispose of unused or expired medications?

Follow local guidelines for medication disposal or use take-back programs.

42. What is the importance of follow-up appointments in medication management?

Follow-up appointments help to monitor progress, manage side effects, and adjust treatment plans.

43. How can people with diabetes prepare for medication changes during different life stages?

Discuss potential changes with healthcare providers and adjust medications as needed for pregnancy, aging, or other life events.

44. How can people with diabetes manage medications during emergencies or natural disasters?

Keep an emergency supply of medications, a list of prescriptions, and a plan for accessing healthcare.

45. What are some signs that diabetes medications may need to be adjusted?

Signs include frequent high or low blood sugar levels, side effects, and changes in health status.

46. How can people with diabetes handle medication intolerance?

Discuss alternative options with healthcare providers and explore different dosages or formulations.

47. What is the role of non insulin injectables in diabetes management?

Noninsulin injectables like GLP-1 receptor agonists and amylin analogs help to regulate blood sugar and manage weight.

48. How can people with diabetes ensure they are taking their medications correctly?

Follow instructions carefully, ask for clarification from healthcare providers, and use medication aids like apps or pillboxes.

49. What are some tips for managing diabetes medications during hospital stays?

Inform hospital staff about all medications, keep a medication list, and monitor blood sugar levels closely.

50. How can people with diabetes stay informed about new medications and treatments?

Follow reputable sources, join diabetes organizations, and discuss new options with healthcare providers.

Conclusion

In conclusion, effective medication management is a cornerstone of diabetes care, crucial for achieving and maintaining optimal blood sugar control, preventing complications, and enhancing overall quality of life. Throughout this chapter, we have explored the diverse landscape of diabetes medications, ranging from insulin therapies to oral medications and non insulin injectables, each tailored to meet the specific needs of individuals based on their type of diabetes and health status.

Understanding how these medications work—whether by enhancing insulin production, improving insulin sensitivity, or managing glucose reabsorption—provides essential insights into their roles in diabetes management. Furthermore, recognizing the potential side effects and interactions associated with each medication type underscores the importance of medication adherence, regular monitoring, and proactive communication with healthcare providers.

Technology, such as insulin pumps and continuous glucose monitors (CGMs), continues to revolutionize diabetes management by offering more precise control and real-time feedback on blood sugar levels. These advancements empower individuals with diabetes to monitor their health more effectively and make timely adjustments to their treatment plans.

Moreover, the integration of medication management strategies into daily routines, travel plans, and emergency preparedness ensures continuity of care and supports long-term health outcomes. Educating oneself about new medications, staying informed about treatment innovations, and actively engaging in shared decision-making with healthcare teams are vital aspects of successful medication management.

As we look to the future, ongoing research and development in diabetes medications hold promise for further improving treatment options and outcomes. By embracing a holistic approach that includes lifestyle modifications, regular physical activity, and stress management alongside medication adherence, individuals with diabetes can achieve sustainable control over their condition and lead healthy, fulfilling lives.

This chapter serves as a comprehensive guide, equipping individuals with diabetes and their caregivers with the knowledge and tools needed to navigate the complexities of medication management effectively. By applying the principles outlined here, individuals can optimize their diabetes care, minimize complications, and strive toward optimal health and well-being.

Bibliography

1. American Diabetes Association. (n.d.). *Insulin basics for diabetes.* https://diabetes.org/health-wellness/medication/insulin-basics
2. American Diabetes Association. (n.d.). *Type 2 diabetes medications.* https://diabetes.org/health-wellness/medication/type-2-medications
3. American Diabetes Association. (n.d.). *Oral and injectable medications for type 2 diabetes.* https://diabetes.org/health-wellness/medication/oral-other-injectable-diabetes-medications
4. Centers for Disease Control and Prevention. (2024, May 15). *Use of new diabetes medicines.* https://www.cdc.gov/diabetes/data-research/research/new-diabetes-medicines.html
5. Mayo Clinic. (2025, May 23). *Diabetes treatment: Medications for type 2 diabetes.* https://www.mayoclinic.org/diseases-conditions/type-2-diabetes/in-depth/diabetes-treatment/art-20051004
6. National Institute of Diabetes and Digestive and Kidney Diseases. (2022, March). *Insulin, medicines, & other diabetes treatments.* https://www.niddk.nih.gov/health-information/diabetes/overview/insulin-medicines-treatments
7. National Institute of Diabetes and Digestive and Kidney Diseases. (2023, October). *Managing diabetes.* https://www.niddk.nih.gov/health-information/diabetes/overview/managing-diabetes
8. World Health Organization. (2024, November 14). *Diabetes.* https://www.who.int/news-room/fact-sheets/detail/diabetes

Emotional and Mental Health in Diabetes

CHAPTER 6

Introduction

This chapter explores the intricate relationship between diabetes and mental well-being, emphasizing that managing diabetes is not solely a physical challenge but also an emotional one. This chapter addresses the profound impact that diabetes can have on mental health, including stress, anxiety, depression, and a condition specific to diabetes known as diabetes distress.

Living with diabetes often involves constant vigilance, strict routines, and the pressure of maintaining blood sugar levels, which can be overwhelming. Diabetes distress manifests in feelings of frustration, burnout, and being overwhelmed. Recognizing these signs is crucial for individuals with diabetes to seek appropriate support and manage their emotional health effectively.

The chapter delves into the strong link between diabetes and depression, highlighting the higher risk faced by individuals managing this chronic condition. It outlines the symptoms of depression, such as persistent sadness and loss of interest in activities, and provides guidance on seeking help from mental health professionals and support groups.

Anxiety is another significant concern for people with diabetes, as it can negatively affect blood sugar levels and overall management. The chapter presents strategies for managing anxiety, including relaxation techniques, cognitive-behavioral therapy (CBT), and medication when necessary.

Therapy and counseling are valuable tools for those struggling with the emotional aspects of diabetes. These resources offer coping strategies, emotional support, and assistance in addressing mental health challenges. Support groups also play a crucial role by providing a sense of community, shared experiences, and emotional backing.

Practical tips for managing stress, practicing mindfulness, and balancing self-care with diabetes management are provided.

The importance of quality sleep, regular physical activity, and a positive outlook are emphasized as vital components of mental health in diabetes management.

Ultimately, this chapter underscores that addressing the emotional and mental health aspects of diabetes is essential for holistic diabetes care. By recognizing and managing these challenges, individuals with diabetes can improve their overall well-being and lead fulfilling lives.

1. How can diabetes impact mental health?

Managing diabetes can lead to stress, anxiety, depression, and diabetes distress.

2. What is diabetes distress?

Diabetes distress refers to the emotional burdens and worries specific to managing diabetes.

3. What are some signs of diabetes distress?

Signs include feelings of frustration, burnout, and being overwhelmed by diabetes management tasks.

4. How can people with diabetes manage stress?

Stress management techniques include exercise, mindfulness practices like meditation or deep breathing, engaging in enjoyable hobbies, and seeking social support from friends, family, or support groups.

▲ **Figure 6.1**
Support group.

5. What is the link between diabetes and depression?

People with diabetes are at a higher risk of depression due to the chronic nature of the condition and its management demands. Additionally, depression can increase the likelihood of poor diabetes management, creating a two-way link where each condition can worsen the other.

6. What are the symptoms of depression in people with diabetes?

Symptoms include persistent sadness, loss of interest or pleasure in activities, changes in appetite or weight, difficulty concentrating, and feelings of hopelessness or guilt. These symptoms may also overlap with diabetes-related fatigue, making it important to seek professional support for an accurate diagnosis.

7. How can people with diabetes seek help for depression?

Speaking with a family doctor is an important first step, as they can assess symptoms and provide referrals to mental health professionals, such as therapists or psychiatrists. Additional support can be found through healthcare providers who specialize in diabetes care, and by participating in support groups.

8. How does anxiety affect people with diabetes?

Anxiety can increase blood sugar levels through the release of stress hormones like cortisol, which affect glucose metabolism. It can also interfere with diabetes management by making it harder to focus on tasks like monitoring blood sugar, adhering to medication schedules, or making healthy lifestyle choices. Common symptoms of anxiety include excessive worry, restlessness, difficulty concentrating, irritability, and physical symptoms like a racing heart, sweating, or shortness of breath.

9. What are some strategies for managing anxiety in people with diabetes?

Strategies include consulting a family physician to discuss symptoms and explore treatment options such as relaxation techniques, cognitive-behavioral therapy (CBT), and medication if necessary. A family physician can also help coordinate care with mental health professionals or diabetes specialists.

10. How can people with diabetes benefit from therapy or counseling?

Therapy can provide valuable coping strategies for managing the emotional and mental health challenges associated with diabetes. It offers a safe space for emotional support, helps individuals develop resilience, and addresses issues such as stress, anxiety, depression, or diabetes distress, ultimately improving both mental health and diabetes management.

11. What is the role of support groups for people with diabetes?

Support groups provide a sense of community where individuals can share experiences, exchange practical advice, and receive emotional support from others who understand the challenges of living with diabetes. They can help reduce feelings of isolation and foster motivation for effective diabetes management.

12. How can people with diabetes manage negative feelings?

People with diabetes can manage negative feelings by acknowledging that diabetes management is challenging and by practicing self-compassion. Seeking support from loved ones or professionals, and focusing on small successes and progress rather than striving for perfection, can help improve emotional well-being and reduce stress.

13. What are some relaxation techniques for people with diabetes?

Some techniques include deep breathing, meditation, progressive muscle relaxation, and yoga.

14. How can people with diabetes balance self-care and diabetes management?

Balancing self-care and diabetes management involves prioritizing activities that promote physical and emotional well-being, such as regular exercise, healthy eating, and relaxation. Setting realistic goals, establishing a consistent routine, and seeking support from healthcare professionals or loved ones when needed can help individuals manage both their health and daily life more effectively.

15. What is the importance of sleep for mental health in diabetes management?

Quality sleep is crucial for mental health as it helps reduce stress, improve mood, and enhance cognitive function. Adequate rest also plays a vital role in regulating blood sugar levels, supporting better diabetes management, and helping individuals cope with the emotional challenges of living with diabetes.

16. How can people with diabetes improve their sleep quality?

People with diabetes can improve sleep quality by maintaining a consistent sleep schedule, creating a calming bedtime routine, and optimizing their sleep environment for comfort. Additionally, addressing any underlying sleep disorders, such as sleep apnea, can further enhance rest and overall health.

17. What is mindfulness and how can it benefit people with diabetes?

Mindfulness is the practice of being fully present and aware of the moment without judgment, which helps reduce stress, improve emotional well-being, and enhance focus. For people with diabetes, mindfulness can improve self-regulation, reduce anxiety related to disease management, and foster a more positive outlook on both their health and daily challenges.

18. How can people with diabetes practice mindfulness?

People with diabetes can practice mindfulness through activities such as meditation, mindful eating, and mindful movement practices like yoga or tai chi. These practices help cultivate awareness, reduce stress, and enhance emotional balance, all of which contribute to better overall diabetes management.

19. What are some common mental health challenges for children with diabetes?

Challenges include coping with the emotional impact of the diagnosis, navigating peer relationships—such as explaining their condition or dealing with misunderstandings—and managing the responsibilities of diabetes care, like monitoring blood sugar levels or adhering to dietary restrictions.

20. How can parents support their children with diabetes?

Parents play a crucial role in recognizing early signs of emotional distress in their children, such as anxiety or withdrawal, and offering initial emotional support. They can provide a safe space for open communication, educate themselves about diabetes, and ensure their child feels understood. Practically, parents can help by creating a supportive home environment, encouraging healthy routines, and seeking guidance from healthcare providers or mental health professionals when necessary. This holistic approach helps the child navigate the emotional challenges of living with diabetes.

21. How can schools support students with diabetes?

Schools can support students with diabetes by creating a supportive environment, providing necessary accommodations (such as allowing time for blood sugar monitoring and insulin administration), and ensuring staff are educated about diabetes management. Additionally, schools can foster an inclusive atmosphere by promoting awareness, addressing any stigma, and offering access to mental health support or counseling when needed. Regular communication between parents, healthcare providers, and school staff can help ensure the student's well-being.

22. What is the role of a diabetes educator in mental health?

Diabetes educators provide education, emotional support, and resources to help manage the psychological aspects of diabetes.

23. How can people with diabetes handle feelings of isolation?

People with diabetes can handle feelings of isolation by connecting with support groups, both in-person and online, where they can share experiences and gain emotional support. Engaging in social activities and maintaining open communication with friends and family can also help reduce feelings of loneliness. It's important to recognize that isolation can sometimes be linked to mental health challenges, so reaching out to a mental health professional for guidance and coping strategies can be a helpful step. Additionally, joining diabetes-focused communities can provide a sense of belonging and reduce feelings of being alone in managing the condition.

24. What are some strategies for coping with diabetes-related fears?

Coping with diabetes-related fears can be helped by educating about the condition, which can reduce uncertainty and empower better decision-making. Developing a clear management plan with the support of healthcare professionals can provide a sense of control and direction. It's also important to acknowledge any emotional or psychological concerns that arise, so seeking professional help, such as from a therapist or counselor, can be a valuable tool in managing anxiety and fear. Engaging in relaxation techniques and mindfulness practices can help reduce stress and provide a calmer perspective on diabetes-related concerns.

25. How can people with diabetes develop a positive outlook?

People with diabetes can develop a positive outlook by focusing on their strengths and what they can control, setting achievable and realistic goals, celebrating both small and big successes, and regularly practicing gratitude. Building a strong support network, engaging in enjoyable activities, and staying informed about their condition can also foster a sense of empowerment and optimism.

26. How can people with diabetes deal with frustration from diabetes management?

To manage frustration from diabetes care, it can help to break tasks into smaller, manageable steps to avoid feeling overwhelmed. Seeking support from friends, family, or diabetes support groups can provide emotional reassurance and practical advice. Practicing stress-relief techniques, such as mindfulness, deep breathing, or physical activity, can also help. Additionally, focusing on progress rather than perfection and giving oneself grace during challenging moments can reduce frustration.

27. What is the impact of chronic stress on diabetes management?

Chronic stress can lead to poor blood sugar control and increased risk of complications. Managing stress through relaxation techniques, regular physical activity, and support from healthcare providers is essential for minimizing these effects and maintaining better overall diabetes management.

28. How can people with diabetes practice self-compassion?

Practice kindness, acknowledge efforts, and avoid self-criticism. Recognizing the challenges of managing diabetes and celebrating progress, rather than focusing on setbacks, can help promote a healthier mindset and reduce feelings of frustration or guilt.

29. What are some benefits of journaling for people with diabetes?

Journaling can help process emotions, track progress, and identify patterns in diabetes management.

30. How can people with diabetes manage the emotional impact of blood sugar fluctuations?

Recognize that fluctuations are a normal part of diabetes management and can occur due to factors like stress, diet, or illness. Use these fluctuations as opportunities to learn and adjust management strategies accordingly. If emotional distress becomes overwhelming or persistent, consider reaching out to a healthcare provider or mental health professional for additional support and guidance.

31. How can people with diabetes build a support network?

People with diabetes can build a strong support network by connecting with healthcare providers, joining diabetes support groups or online communities, and involving friends and family in their care.

32. How can people with diabetes manage the emotional impact of complications?

People with diabetes can manage the emotional impact of complications by seeking support from loved ones, support groups, or mental health professionals. Focusing on preventive measures, maintaining a positive outlook, and working closely with healthcare providers to address complications can help reduce anxiety and improve overall well-being.

33. What are some strategies for maintaining motivation in diabetes management?

To maintain motivation, set realistic and achievable goals, regularly track progress, and celebrate small successes along the way. Rewarding accomplishments and seeking support from healthcare providers, family, or diabetes communities can provide encouragement and help sustain long-term commitment to managing the condition.

34. How can people with diabetes address diabetes-related stigma?

People with diabetes can address stigma by educating others about the condition, advocating for their needs, and seeking support from those who share similar experiences or understand the challenges of living with diabetes.

35. What is the role of family in supporting diabetes management?

Family can provide emotional support, help with management tasks, and create a positive home environment.

36. How can people with diabetes handle burnout?

To handle burnout, people with diabetes can take regular breaks, seek emotional support from loved ones or professionals, and adjust their management strategies to reduce stress. Focusing on self-care activities such as rest, relaxation, and hobbies can help recharge energy and maintain a balanced approach to managing the condition.

37. How can humor benefit people with diabetes?

Humor can reduce stress, improve mood, and provide a positive perspective on managing diabetes.

38. How can people with diabetes find balance between diabetes management and daily life?

People with diabetes can find balance by integrating management tasks, like monitoring blood sugar or taking medications, into their daily routines to make them feel less disruptive. Setting clear boundaries to avoid overcommitment and prioritizing enjoyable activities can help maintain a healthy balance between managing the condition and enjoying life.

39. How can people with diabetes manage frustration with their condition?

People with diabetes can manage frustration by recognizing it as a natural response to the challenges of the condition and allowing themselves to feel it without guilt. Seeking support from healthcare providers or support groups, adjusting management approaches, and focusing on small, achievable goals can help reduce frustration and maintain a sense of control.

40. What are some tips for managing diabetes in social situations?

Prepare in advance by bringing necessary supplies and snacks, communicate openly with friends and family about diabetes needs, and adapt management strategies to fit the situation while maintaining health priorities.

41. How can people with diabetes deal with feelings of anger or resentment?

Practice stress-relief techniques, seek counseling, and focus on aspects of life that bring joy and fulfillment.

42. How can people with diabetes improve their self-esteem?

Focus on strengths, set and achieve personal goals, and practice self-acceptance.

43. What is the role of spirituality in managing diabetes?

Spirituality can provide comfort, strength, and a sense of purpose, helping to cope with the challenges of diabetes.

44. How can pets help people with diabetes?

Pets can provide companionship, reduce stress, and encourage physical activity.

45. How can people with diabetes use art or creative activities for emotional well-being?

Art and creative activities offer a therapeutic outlet for self-expression, helping to reduce stress, improve mood, and foster emotional resilience. Engaging in these activities can provide a sense of accomplishment, enhance relaxation, and serve as a valuable tool for coping with the emotional challenges of living with diabetes.

46. How can people with diabetes develop resilience?

People with diabetes can develop resilience by building a strong support network, focusing on problem-solving skills to manage challenges, and maintaining a positive outlook despite setbacks. Embracing flexibility, learning from experiences, and practicing self-compassion also contribute to greater emotional strength and the ability to cope with the ups and downs of managing the condition.

47. What are some benefits of nature for people with diabetes?

Spending time in nature can reduce stress, improve mood, and promote physical activity.

48. How can people with diabetes practice gratitude?

Keep a gratitude journal, express appreciation to others, and focus on positive aspects of life.

49. How can people with diabetes find balance between independence and seeking help?

Strive for a balance by learning to identify when self-management is sufficient and when additional support is necessary. This may involve recognizing signs of stress or difficulty in managing tasks alone. It's important to communicate clearly with healthcare providers, family members, or support groups about needs and limitations. While maintaining autonomy is important, asking for help when needed can prevent burnout and ensure better overall health management.

50. How can people with diabetes maintain hope and optimism?

Set realistic goals, celebrate progress, seek positive role models, and stay connected with supportive communities.

Conclusion

This chapter has explored the profound intersection of emotional and mental health with diabetes management, highlighting the significant impact these factors have on individuals' well-being. Managing diabetes goes beyond physical health; it entails navigating complex emotional challenges, including stress, anxiety, depression, and diabetes-specific distress.

Throughout this chapter, we have underscored the importance of recognizing and addressing these emotional aspects. Diabetes distress, characterized by feelings of frustration, burnout, and overwhelm due to the demands of diabetes management, is a common experience among individuals with diabetes. Understanding its signs and seeking appropriate support are crucial steps toward mitigating its impact.

The chapter also delved into the heightened risk of depression and anxiety among people with diabetes, emphasizing the need for proactive management and seeking professional help when necessary. Strategies such as relaxation techniques, mindfulness practices, and therapy have been highlighted as effective tools for managing these conditions and promoting overall well-being.

Support groups were identified as invaluable resources for providing community, understanding, and emotional support. They offer a platform for individuals to share experiences, gain insights, and receive encouragement from peers facing similar challenges.

Practical tips for balancing self-care with diabetes management, improving sleep quality, and fostering a positive mindset have been provided to empower individuals in their journey toward emotional resilience. The integration of these strategies not only enhances mental health but also contributes to better diabetes outcomes and overall quality of life.

Ultimately, this chapter reinforces the holistic approach needed in diabetes care—one that addresses both the physical and emotional dimensions of the condition. By acknowledging and addressing emotional health alongside medical management, individuals with diabetes can cultivate resilience, improve coping strategies, and achieve a balanced life despite the challenges they face. Through education, support, and proactive self-care, the path to managing diabetes becomes more manageable and sustainable for long-term health and well-being.

Bibliography

1. American Diabetes Association. (n.d.). *Mental health.* https://diabetes.org/health-wellness/mental-health
2. Centers for Disease Control and Prevention. (2024, May 15). *Diabetes and mental health.* https://www.cdc.gov/diabetes/living-with/mental-health.html
3. Diabetes Canada. (n.d.). *How diabetes can affect your mental health.* https://www.diabetes.ca/archive/managing-my-diabetes---archive/preventing-complications/mental-health
4. Robinson, D. J., Hanson, K., Jain, A. B., Kichler, J. C., Mehta, G., Melamed, O. C., Vallis, M., Bajaj, H. S., Barnes, T., Gilbert, J., Honshorst, K., Houlden, R., Kim, J., Lewis, J., MacDonald, B., MacKay, D., Mansell, K., Rabi, D., Sherifali, D., & Senior, P. (2023). Diabetes and mental health. *Canadian Journal of Diabetes, 47*(4), 308–344. https://doi.org/10.1016/j.jcjd.2023.04.009
5. National Institute of Mental Health. (2024, December). *Depression.* https://www.nimh.nih.gov/health/topics/depression
6. National Institute of Mental Health. (2024, December). *Anxiety disorders.* https://www.nimh.nih.gov/health/topics/anxiety-disorders
7. Morales-Brown, L. A., Perez Algorta, G., & Salifu, Y. (2024). Understanding e xperiences of d iabetes d istress: A s ystematic r eview and t hematic s ynthesis. *Journal of D iabetes R esearch, 2024*, 3946553. https://doi.org/10.1155/2024/3946553
8. American Psychiatric Association. (2013). *Diagnostic and statistical manual of mental disorders* (5th ed.). https://doi.org/10.1176/appi.books.9780890425596
9. Akhaury, K., & Chaware, S. (2022). Relation b etween d iabetes and p sychiatric disorders. *Cureus, 14*(10), e30733. https://doi.org/10.7759/cureus.30733

CHAPTER 7

Diabetes and Travel

Introduction

Traveling can be an enriching experience for anyone, including those living with diabetes. However, managing diabetes while on the move requires careful planning and preparation to ensure that blood sugar levels remain stable and that any potential health issues are addressed promptly. This chapter provides a comprehensive guide to help individuals with diabetes navigate the unique challenges that come with traveling, whether for leisure, work, or family visits.

This chapter begins by emphasizing the importance of thorough preparation before embarking on a trip. It covers essential steps such as packing the necessary diabetes supplies, including medications, blood sugar monitors, insulin, and emergency glucagon. Travelers are advised to keep these supplies in their carry-on luggage to prevent any disruptions due to lost or delayed checked baggage. The importance of carrying a prescription list and knowing where to find medical care in the travel destination is also highlighted.

Managing diabetes medications while traveling can be complex, especially when crossing time zones. The chapter provides practical tips for adjusting insulin timings and maintaining a consistent medication schedule to avoid any adverse effects on blood sugar levels. It also offers advice on handling meal planning, ensuring access to healthy food options, and staying hydrated, which are crucial aspects of diabetes management during travel.

The chapter addresses various modes of travel, including flying, driving, and taking cruises, with specific tips tailored to each. For instance, it advises on how to inform TSA about diabetes supplies when flying and how to manage blood sugar levels during long flights. For road trips, it suggests keeping supplies within reach and taking regular breaks to check blood sugar.

Additionally, the chapter explores strategies for managing stress, staying active, and maintaining a healthy routine while on the road. It also covers more specific scenarios, such as traveling to different climates, handling time zone changes, and managing dietary restrictions. By following the guidance provided in this chapter, individuals with diabetes can enjoy their travel experiences while keeping their health in check, ensuring a safe and enjoyable journey.

1. How can people with diabetes prepare for travel?

Plan ahead by packing sufficient supplies, carrying a prescription list, and knowing where to find medical care.

2. What diabetes supplies should be packed for travel?

Include medications, blood sugar monitors, insulin, snacks, emergency glucagon, and extra test strips. It's also important to include a travel log for blood sugar readings, a syringe or insulin pen, a cooler or case for insulin, and a letter from a healthcare provider explaining the condition and necessary treatments, especially for air travel or international trips.

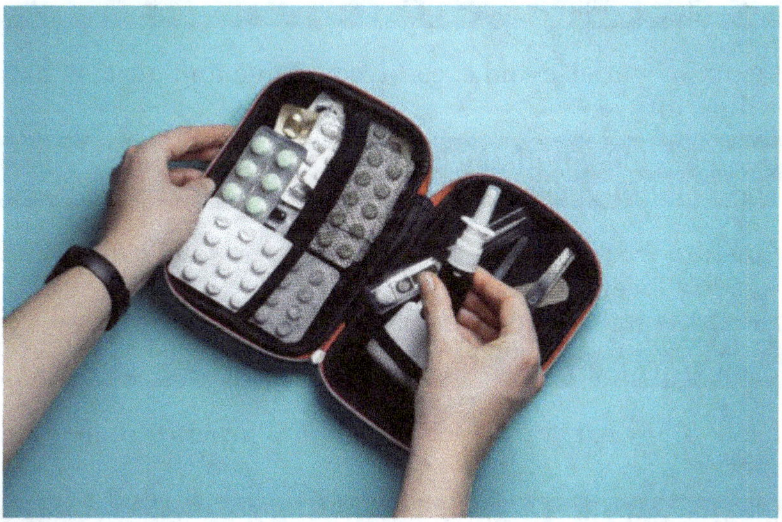

▲ Figure 7.1
Travel first-aid kit.

3. How can people with diabetes manage their medications while traveling?

Keep medications in carry-on luggage, adjust timing for time zone changes, and pack a backup supply of essentials, including extra medications, testing supplies, fast-acting glucose for low blood sugar, and prescription copies in case replacements are needed.

4. What are some tips for flying with diabetes?

Inform the Transportation Security Administration (TSA) or relevant airport security agency about diabetes supplies, keep medications accessible, and adjust insulin timing if crossing time zones.

5. How can people with diabetes handle meal planning during travel?

Research restaurant options, pack healthy snacks, and maintain regular meal times.

6. What should people with diabetes do if they experience low blood sugar during travel?

Carry fast-acting glucose, monitor blood sugar frequently, and have a plan for managing lows.

7. How can people with diabetes stay hydrated while traveling?

Drink plenty of water, avoid excessive caffeine and alcohol, and monitor fluid intake.

8. What precautions should people with diabetes take when traveling to different climates?

Protect insulin from extreme temperatures, stay hydrated, and adjust activity levels accordingly.

9. How can people with diabetes manage stress while traveling?

Plan ahead, allow extra time for travel, and practice relaxation techniques.

10. What are some tips for traveling by car with diabetes?

Keep supplies within reach, take regular breaks to check blood sugar, and stay hydrated.

11. How can people with diabetes manage time zone changes?

Adjust medication timing gradually, monitor blood sugar closely, and consult a healthcare provider if needed.

12. What should people with diabetes consider when traveling internationally?

Research healthcare facilities, learn key phrases in the local language, understand medication regulations, and ensure travel insurance covers diabetes-related needs.

13. How can people with diabetes handle changes in routine during travel?

Maintain consistent meal and medication schedules as much as possible. If unexpected changes occur, like delayed meals or altered plans, carry snacks for emergencies, adjust medication timing with a healthcare provider's guidance, and use reminders to stay on track.

14. What are some tips for eating out while traveling?

When eating out while traveling, choose healthier options like grilled or baked dishes instead of fried, and opt for vegetables or salads as sides. Control portion sizes by sharing meals or asking for half portions and monitor blood sugar levels after meals to assess how different foods affect your body. It's also helpful to carry snacks in case healthier choices aren't available.

15. How can people with diabetes find healthy food options in unfamiliar places?

Research in advance, look for grocery stores instead of relying on fast food, and pack portable, healthy snacks.

16. How can people with diabetes ensure they have enough supplies for longer trips?

Pack extra supplies, carry additional prescriptions, refill medications before leaving, and research where to access supplies at your destination.

17. What are some ways to stay active while traveling?

Walk to explore new places, use hotel gyms, and engage in outdoor activities.

18. How can people with diabetes manage their condition during a cruise?

Inform the cruise line of dietary needs in advance, pack sufficient diabetes supplies, monitor blood sugar regularly, stay active with onboard activities, and stay hydrated. Additionally, prepare for emergencies by carrying fast-acting glucose, having a medical ID, and knowing where to access medical care on the ship or at ports.

19. What should people with diabetes do if they need medical care while traveling?

Know how to contact local healthcare providers, carry health insurance information, and have a medical alert ID.

20. How can people with diabetes manage their condition at theme parks?

Plan meals and medication around activities, stay hydrated, and take breaks to check blood sugar. Carry fast-acting glucose and snacks for emergencies. Be mindful of physical activity and schedule rest to avoid exhaustion.

21. How can people with diabetes handle long flights?

Move around regularly, stay hydrated, and keep diabetes supplies accessible.

22. What are some tips for traveling with children with diabetes?

Pack extra supplies like medications and snacks. Involve children in planning, and ensure they stick to their diabetes routine. Keep them engaged with activities and take breaks to check blood sugar. Plan for meals and insulin around time zones.

23. How can people with diabetes manage their condition during road trips?

Pack diabetic-friendly snacks and research dining options. Take breaks to check blood sugar and ensure timely medication. Stay hydrated with water and avoid sugary drinks. Bring extra supplies, including insulin and glucose tablets.

24. How can people with diabetes stay safe while participating in travel activities?

Monitor blood sugar levels before, during, and after activities to prevent hypoglycemia or hyperglycemia. Stay hydrated with water or sugar-free beverages, especially in hot or physically demanding environments. Carry snacks like glucose tablets or low-carb options in case of sudden drops in blood sugar. Pay attention to signs of fatigue, dizziness, or weakness, which may indicate the need for a break or blood sugar adjustment. It's important to inform travel companions or activity organizers about the condition and emergency procedures. When engaging in physical activities like hiking, swimming, or sports, consider wearing a medical alert bracelet for added safety.

25. What should people with diabetes do if they lose their supplies while traveling?

Stay calm and have a backup plan in place. Know the location of nearby pharmacies, clinics, or medical supply stores where you can get replacements. It's also a good idea to have a list of prescriptions and dosages and carry a doctor's letter explaining the condition and medication needs. If traveling internationally, research in advance how to access diabetes supplies in destination country. Additionally, consider carrying extra medication and supplies in multiple bags or with travel companions to reduce the risk of losing everything at once.

26. How can people with diabetes manage their condition during active vacations?

Monitor blood sugar levels more frequently, especially before, during, and after physical activities. Adjust insulin doses as needed to account for increased exercise or changes in food intake. Stay hydrated by drinking water regularly and be mindful of the effects of physical activity on blood sugar levels. It's also important to have quick-acting carbohydrates on hand in case of hypoglycemia and inform activity guides or travel companions about the condition. Planning meals and snacks around activity schedules is crucial to maintain stable blood sugar levels throughout the day.

27. What are some tips for visiting high-altitude destinations with diabetes?

Monitor blood sugar more closely, as high altitudes can affect insulin sensitivity and blood sugar levels. Stay hydrated by drinking plenty of water, as dehydration is common at high elevations and can impact blood sugar control. Adjust activity levels to avoid overexertion, as physical exertion can cause blood sugar fluctuations at higher altitudes. Be mindful of altitude-related symptoms like dizziness or fatigue, which could be signs of low blood sugar. Also, take time to acclimate to the altitude gradually to minimize stress on the body. Carry extra supplies, as access to medical care or diabetes supplies might be limited in remote areas.

28. How can people with diabetes manage their condition during holidays and celebrations?

Plan meals, monitor blood sugar, and watch portion sizes. Stay active, communicate dietary needs, and manage stress. Get adequate rest to stabilize blood sugar levels.

29. How can people with diabetes handle jet lag?

Gradually adjust to new time zones, stay hydrated, and monitor blood sugar. Use devices like CGMs or insulin pumps to track levels. Consult a healthcare provider for travel-specific advice.

30. How can people with diabetes find support while traveling?

Connect with local diabetes groups, use online resources, and communicate with travel companions.

31. What are some benefits of travel for people with diabetes?

Travel can reduce stress, provide new experiences, and promote physical activity.

32. How can people with diabetes ensure they have access to healthy foods while traveling?

Research dining options in advance and use grocery delivery services when possible. Pack portable snacks like nuts, fruits, and protein bars to manage blood sugar between meals.

33. What should people with diabetes consider when traveling to remote locations?

Ensure access to medical care by researching local healthcare facilities. Pack extra supplies, including insulin, test strips, and medications. Be prepared for emergencies by having a list of emergency contacts and a medical ID with your diabetes information.

34. How can people with diabetes manage their condition during outdoor adventures?

Monitor blood sugar before and after activities. Carry insulin, snacks, testing supplies, and a glucagon kit. Stay hydrated, plan meals around activity levels, and adjust for the impact of exercise on blood sugar. Inform companions and wear a medical ID. Take breaks during intense activities to check blood sugar and adjust as needed.

35. How can people with diabetes handle travel-related fatigue?

Get adequate rest, stay hydrated, and manage blood sugar levels. Consider adjusting activity levels to prevent exhaustion and listen to your body for signs of fatigue.

36. How can people with diabetes maintain a healthy routine during travel?

Stick to meal and medication schedules, stay active, and monitor blood sugar regularly. If possible, plan for breaks during the day to check blood sugar and manage routine.

37. What are some tips for managing diabetes during camping trips?

Keep supplies safe from extreme temperatures, plan meals in advance, and monitor blood sugar. Make sure insulin and medications are stored at the proper temperatures and have a reliable way to check blood sugar while outdoors.

38. How can people with diabetes prepare for travel emergencies?

Have a backup plan in case of lost supplies or emergencies, carry emergency contact information, and know how to access local healthcare. It's also helpful to carry a medical ID and a list of medications.

39. How can people with diabetes manage their condition during busy travel schedules?

Plan ahead, pack convenient snacks, and take regular breaks to check blood sugar. Keep diabetes management tools within easy reach and adjust your routine to accommodate travel delays or unexpected changes.

40. What should people with diabetes consider when booking accommodations?

Look for accommodations with kitchen facilities, easy access to food stores, and comfortable environments. Ensure that accommodations can meet specific needs, such as access to refrigeration for medications.

41. How can people with diabetes handle dietary restrictions while traveling?

Communicate needs to restaurants, pack suitable snacks, and research options in advance.

42. What are some ways to stay active on a beach vacation for people with diabetes?

Swim, walk along the beach, and participate in water sports. Remember to monitor blood sugar before, during, and after activities to prevent fluctuations due to physical exertion.

43. How can people with diabetes prepare for long-distance travel?

Plan for meal and medication timing, pack extra supplies, and stay flexible.

44. How can people with diabetes manage their condition during family visits?

Communicate dietary needs, plan meals in advance, and monitor blood sugar.

45. How can people with diabetes ensure they get enough sleep while traveling?

Maintain a regular sleep schedule, create a comfortable sleep environment, and manage jet lag.

46. What are some tips for hiking trips for people with diabetes?

Pack extra supplies, monitor blood sugar, and stay hydrated. Be prepared for sudden changes in activity levels and check blood sugar frequently, especially if hiking at higher altitudes.

47. How can people with diabetes stay organized while traveling?

Use packing lists, keep supplies in a designated bag, and have a travel itinerary. Make sure essential medical supplies, snacks, and medications are easily accessible throughout the trip.

48. How can people with diabetes manage their condition during cruises?

Inform the cruise line, plan for meals, and monitor blood sugar regularly.

49. What are some ways to handle travel delays for people with diabetes?

Carry extra snacks, keep supplies accessible, and stay flexible with plans. Have a backup plan for meals and medication adjustments during delays and ensure access to adequate insulin supplies or medication in case of prolonged delays.

50. How can people with diabetes enjoy their travel experiences?

Plan ahead, stay flexible, and focus on enjoying new experiences and activities.

Conclusion

This chapter discussing diabetes and travel underscores the importance of meticulous planning and vigilant management to ensure a safe and enjoyable travel experience for individuals with diabetes. Traveling can pose unique challenges for diabetes management, but with the right strategies and precautions, these challenges can be effectively navigated.

Preparation is key, and this chapter provides comprehensive advice on packing essential diabetes supplies, including medications, blood sugar monitors, insulin, and emergency glucagon. Keeping these supplies in carry-on luggage ensures they are readily accessible and prevents complications from lost or delayed checked baggage. Having a prescription list and knowing where to find medical care at the travel destination further safeguards against potential health issues.

Managing medications and blood sugar levels across different time zones and varying travel schedules can be complex. Practical tips on adjusting insulin timings and maintaining a consistent medication routine help mitigate these complexities. The chapter also emphasizes the importance of meal planning, staying hydrated, and having healthy snacks on hand, which are crucial for maintaining stable blood sugar levels during travel.

Specific travel modes, such as flying, driving, and cruising, each come with unique considerations. Guidance on navigating airport security with diabetes supplies, managing blood sugar on long flights, and ensuring regular monitoring during road trips is provided to ensure smooth travel experiences. The chapter also offers strategies for staying active, managing stress, and adhering to a healthy routine while traveling, which are vital for overall well-being.

Additional scenarios, such as traveling to different climates, handling time zone changes, and managing dietary restrictions, are addressed with practical solutions. By following the tips and recommendations in this chapter, individuals with diabetes can effectively manage their condition while enjoying new experiences and destinations.

In conclusion, with careful planning and proactive management, people with diabetes can travel safely and confidently. This chapter equips travelers with the knowledge and tools needed to navigate their journeys successfully, ensuring both their health and enjoyment are prioritized.

Bibliography

1. NMC Health. (2021, July 2). *People with diabetes: Traveling safe this summer.* https://www.mynmchealth.org/traveling-safe-this-summer/
2. Centers for Disease Control and Prevention. (2024, May 15). *Tips for traveling with diabetes.* https://www.cdc.gov/diabetes/about/tips-for-traveling-with-diabetes.html
3. Diabetes Canada. (n.d.). *Air travel.* https://www.diabetes.ca/learn-about-diabetes/your-rights/air-travel
4. International Diabetes Federation. (2020, December 21). *Traveling and diabetes.* IDF Europe. https://idf.org/europe/media/uploads/sites/2/2023/06/AwarenessPaper-DiabetesAndTravelling-21Dec2020_compressed.pdf
5. Mayo Clinic. (2020, April 17). *Diabetes and travel: Planning is key.* https://www.augustahealth.com/answer/diabetes-and-travel-planning-is-key/
6. Johns Hopkins Medicine. (n.d.). *Traveling with diabetes.* https://www.hopkinsmedicine.org/health/conditions-and-diseases/diabetes/traveling-with-diabetes
7. WebMD. (2024, March 14). *On the road: Traveling with diabetes.* https://www.webmd.com/diabetes/traveling
8. World Health Organization. (n.d.). *Travel and health.* https://www.who.int/health-topics/travel-and-health#tab=tab_1

CHAPTER 8

Diabetes and Sleep

Introduction

This chapter delves into the critical yet often overlooked relationship between diabetes and sleep health. Sleep disturbances are common among individuals with diabetes, primarily due to fluctuating blood sugar levels, neuropathy pain, and frequent urination. The chapter emphasizes the vital importance of quality sleep for people with diabetes, highlighting how it supports blood sugar control, reduces stress, and improves overall health.

Improving sleep quality can significantly impact diabetes management. This chapter offers practical strategies for enhancing sleep, such as maintaining a regular sleep schedule, creating a relaxing bedtime routine, and effectively managing blood sugar levels. It addresses common sleep disorders associated with diabetes, including sleep apnea, restless legs syndrome, and insomnia, providing guidance on their management.

The chapter also discusses specific issues like preventing nighttime hypoglycemia, managing nocturia, and the role of diet and physical activity in promoting better sleep. Creating a sleep-friendly environment, stress management, and dealing with insomnia are covered in detail, offering actionable tips for people with diabetes to improve their sleep hygiene and overall well-being.

Special attention is given to the impact of poor sleep on blood sugar levels, revealing how inadequate rest can lead to higher blood sugar levels and increased insulin resistance. Techniques for tracking sleep patterns, the benefits of napping, and managing pain that disrupts sleep are explored to help individuals maintain a balanced lifestyle.

Furthermore, the chapter touches on the role of melatonin, handling nighttime blood sugar testing, and managing shift work and irregular sleep schedules. Natural remedies for improving sleep, avoiding screen time before bed, and relaxation techniques are suggested to promote better sleep habits.

By emphasizing the interconnectedness of sleep and diabetes management, Chapter 8 provides a comprehensive guide for people with diabetes to enhance their sleep quality, ultimately leading to better health outcomes. Through practical advice and strategies, this chapter empowers individuals to take control of their sleep health, fostering a holistic approach to diabetes management.

▲ Figure 8.1
The importance of sleep.

1. How does diabetes affect sleep?

Diabetes can cause sleep disturbances due to fluctuating blood sugar levels, neuropathy pain, and frequent urination.

2. What is the importance of good sleep for people with diabetes?

Quality sleep supports blood sugar control, reduces stress, and improves overall health.

3. How can people with diabetes improve their sleep quality?

Maintain a regular sleep schedule, create a relaxing bedtime routine, and manage blood sugar levels.

4. What are common sleep disorders associated with diabetes?

Sleep apnea, restless legs syndrome, and insomnia are common among people with diabetes.

5. How can people with diabetes manage sleep apnea?

Use a CPAP machine as prescribed to improve sleep quality. Maintain a healthy weight, as it can reduce symptoms of both sleep apnea and diabetes. Avoid alcohol and sedatives before bedtime, as they can relax throat muscles and worsen apnea. Schedule regular follow-ups or sleep studies to monitor and adjust treatment as needed. Managing blood sugar effectively can also help improve overall sleep quality.

6. What is restless legs syndrome and how can it be managed?

Restless legs syndrome (RLS) is a condition characterized by uncomfortable sensations in the legs, often accompanied by an irresistible urge to move them, especially at night. Management includes medications, maintaining a regular sleep schedule, avoiding caffeine or alcohol, staying active, and practicing relaxation techniques.

7. How can people with diabetes prevent nighttime hypoglycemia?

To prevent nighttime hypoglycemia, consider having a balanced bedtime snack, monitoring blood sugar levels before bed, and, if necessary, adjusting insulin doses under the guidance of a healthcare provider.

8. How can people with diabetes manage nocturia (frequent urination at night)?

Limit fluids before bed, especially caffeine and alcohol. Manage blood sugar levels to reduce excess urination caused by hyperglycemia. Address potential contributing factors such as urinary tract infections, bladder conditions, or sleep apnea with guidance from a healthcare provider.

9. What role does diet play in sleep quality for people with diabetes?

A balanced diet helps stabilize blood sugar levels, which can improve sleep.

10. How can physical activity improve s leep for people with diabetes?

Regular exercise promotes better sleep quality and overall health.

11. What are some tips for creating a sleep-friendly environment?

Keep the bedroom cool, dark, and quiet, and use a comfortable mattress and pillows.

12. How can stress management improve sleep for people with diabetes?

Techniques like mindfulness, deep breathing, and relaxation exercises can reduce stress and improve sleep.

13. How can people with diabetes handle insomnia?

Establish a consistent sleep routine, limit screen time before bed, and avoid caffeine and heavy meals in the evening. Monitor blood sugar levels to address fluctuations that might disrupt sleep. If insomnia persists, consider Cognitive Behavioral Therapy for Insomnia (CBT-I) to improve sleep quality and address underlying causes.

14. What is the impact of poor sleep on blood sugar levels?

Poor sleep can lead to higher blood sugar levels and increased insulin resistance.

15. How can people with diabetes track their sleep patterns?

Use sleep journals, apps, or wearable devices to monitor sleep quality and patterns.

16. What are some benefits of napping for people with diabetes?

Short naps, ideally 20–30 minutes, can improve alertness, mood, and energy levels and are less likely to disrupt nighttime sleep compared to longer naps. Keeping naps brief helps minimize the risk of grogginess or sleep disturbances later in the day.

17. How can people with diabetes manage pain that disrupts sleep?

Address pain with medications, physical therapy, or lifestyle changes to improve sleep. It's important to consult with healthcare providers to determine the most appropriate treatment plan for individual needs.

18. How can people with diabetes establish a bedtime routine?

Develop a relaxing routine with activities like reading, taking a warm bath, or practicing relaxation techniques. Consistency is key, so try to follow the same routine each night to help signal to the body that it's time to wind down for sleep.

19. What is the role of melatonin in sleep for people with diabetes?

Melatonin supplements may help regulate sleep but consult a healthcare provider before use.

20. How can people with diabetes handle nighttime blood sugar testing?

Keep testing supplies nearby and establish a routine to minimize sleep disruption.

21. How can people with diabetes manage shift work and irregular sleep schedules?

Use blackout curtains, maintain a consistent routine, and adjust meal and medication timing.

22. What are some natural remedies for improving sleep for people with diabetes?

Herbal teas, aromatherapy, and magnesium supplements may promote relaxation and better sleep. However, it's important to consult with a healthcare provider before trying any natural remedies to ensure they are safe and won't interfere with diabetes management or other medications.

23. How can people with diabetes avoid screen time before bed?

Set a digital curfew at least 30 minutes to an hour before bed, use blue light filters, and engage in non-screen activities like reading or relaxation techniques to prepare for sleep.

24. What are some relaxation techniques to improve sleep?

Techniques include progressive muscle relaxation, deep breathing exercises, and guided imagery.

25. How can people with diabetes ensure a consistent sleep schedule?

Aim for a consistent sleep-wake schedule, waking up at the same time every day, even on weekends. Going to bed at a similar time each night, based on routine and feelings of tiredness, can also improve sleep quality.

26. How can people with diabetes manage bedtime hunger?

Have a light, healthy snack that includes protein and fiber.

27. How can people with diabetes deal with anxiety that affects sleep?

Practice stress-reduction techniques, seek therapy, and avoid stimulating activities before bed.

28. How can people with diabetes improve sleep hygiene?

Maintain a regular sleep schedule, create a relaxing bedtime routine, and make the sleep environment comfortable.

29. What are some tips for managing blood sugar levels during sleep?

Monitor blood sugar before bed, adjust medications as needed (in consultation with a healthcare provider), and maintain a balanced diet.

30. How can people with diabetes handle nighttime emergencies?

Keep glucose tablets or snacks nearby and have a plan for managing low blood sugar. If living alone, keep a phone nearby to contact someone for help if needed.

31. How can caffeine affect sleep for people with diabetes?

Caffeine can disrupt sleep, so limit intake, especially in the afternoon and evening.

32. How can people with diabetes manage sleep disruptions caused by medications?

Discuss with a healthcare provider to adjust timing or dosage and explore alternative treatments if needed.

33. How can people with diabetes handle travel-related sleep disruptions?

Adjust to new time zones gradually, maintain a sleep routine, and manage blood sugar levels.

34. How can people with diabetes ensure they get enough sleep?

Prioritize sleep by sticking to a consistent sleep schedule, avoiding stimulants like caffeine in the evening, and creating a relaxing bedtime routine. Address underlying health issues, such as sleep apnea, restless legs syndrome, or neuropathy, with guidance from a healthcare provider, as these conditions may interfere with sleep.

35. What are some benefits of good sleep for people with diabetes?

Good sleep improves blood sugar control, reduces stress, and enhances overall health and well-being.

36. How can people with diabetes manage nightmares or vivid dreams?

Address blood sugar fluctuations by monitoring levels before bed and adjusting insulin or medication as needed. Reduce stress with relaxation techniques and create a calming bedtime routine to promote restful sleep. If vivid dreams or nightmares persist, consult a healthcare provider to ensure blood sugar levels are stable overnight.

37. What are some signs of sleep apnea for people with diabetes?

Signs include loud snoring, choking or gasping during sleep, and daytime sleepiness.

38. How can people with diabetes get diagnosed with sleep apnea?

Consult a healthcare provider for a sleep study or home sleep apnea test.

39. How can weight management improve sleep for people with diabetes?

Maintaining a healthy weight can reduce sleep apnea risk and improve overall sleep quality.

40. How can people with diabetes handle sleep disruptions from neuropathy pain?

Use pain management techniques, medications, and consult a healthcare provider for treatment options.

41. What are some tips for parents of children with diabetes to ensure good sleep?

Establish a consistent bedtime routine with calming activities, monitor blood sugar levels before bed, and ensure they are within a safe range to prevent nighttime hypoglycemia or hyperglycemia. Create a supportive sleep environment by reducing screen time, ensuring the room is quiet and dark, and using a comfortable mattress. If needed, consult with a healthcare provider to adjust insulin or medications to improve sleep quality.

42. How can people with diabetes manage restless sleep?

Monitor and manage blood sugar levels, as fluctuations can cause restless sleep. Address any underlying conditions like sleep apnea or restless legs syndrome. Practice relaxation techniques, such as deep breathing or meditation, and maintain good sleep hygiene by creating a consistent bedtime routine, limiting caffeine, and ensuring a comfortable sleep environment. If restless sleep persists, consult a healthcare provider for further evaluation and possible treatment options.

43. How can people with diabetes handle sleep disruptions during pregnancy?

Adjust sleep positions, manage blood sugar levels, and consult a healthcare provider for guidance.

44. How can people with diabetes manage seasonal changes that affect sleep?

Adjust routines to accommodate changes in daylight and maintain a consistent sleep schedule.

45. How can pets affect sleep for people with diabetes?

Pets can be comforting but may disrupt sleep; create boundaries to ensure restful sleep.

46. How can people with diabetes manage sleep disruptions from caregiving responsibilities?

Seek support from family or professionals, prioritize self-care, and create a flexible sleep routine that accommodates caregiving duties.

47. How can people with diabetes handle sleep disruptions from environmental factors?

Use earplugs, blackout curtains, and maintain a comfortable sleep temperature.

48. How can people with diabetes benefit from mindfulness for better sleep?

Mindfulness practices can reduce stress and promote relaxation, improving sleep quality.

49. What are some signs that people with diabetes need professional help for sleep issues?

Persistent sleep problems, severe daytime sleepiness, and difficulty managing blood sugar levels.

50. How can people with diabetes celebrate good sleep habits?

Celebrate healthy sleep routines and acknowledge the positive impact on overall health.

Conclusion

This chapter on diabetes and sleep highlights the intricate relationship between sleep health and diabetes management, underscoring the profound impact that sleep can have on overall well-being. Poor sleep can exacerbate diabetes symptoms, leading to fluctuating blood sugar levels, increased insulin resistance, and heightened stress, which collectively undermine effective diabetes management. Conversely, quality sleep supports better blood sugar control, reduces stress, and enhances general health, emphasizing its critical role in the lives of those managing diabetes.

The chapter provides a comprehensive guide to improving sleep quality for people with diabetes. Key strategies include maintaining a consistent sleep schedule, creating a relaxing bedtime routine, and managing blood sugar levels effectively. By addressing common sleep disorders such as sleep apnea, restless legs syndrome, and insomnia, the chapter offers practical solutions and treatments to alleviate these conditions and promote better sleep.

Understanding the interplay between diet, physical activity, and sleep is crucial. A balanced diet and regular exercise are highlighted as vital components that can stabilize blood sugar levels and contribute to improved sleep quality. Additionally, creating a sleep-friendly environment, managing stress through relaxation techniques, and establishing a bedtime routine are essential practices for fostering better sleep hygiene.

The chapter also addresses specific challenges like preventing nighttime hypoglycemia, managing nocturia, and dealing with pain that disrupts sleep, offering tailored advice to mitigate these issues. It explores the role of melatonin, the importance of tracking sleep patterns, and the potential benefits of napping, providing a well-rounded approach to sleep management.

Ultimately, this chapter empowers individuals with diabetes to take control of their sleep health through practical advice and strategies. By prioritizing and improving sleep, people with diabetes can achieve better health outcomes, enhance their quality of life, and manage their condition more effectively. This holistic approach underscores the importance of integrating good sleep habits into daily diabetes management routines, paving the way for improved overall well-being.

Bibliography

1. Henson, J., Covenant, A., Hall, A. P., Herring, L., Rowlands, A. V., Yates, T., & Davies, M. J. (2024). Waking up to the importance of sleep in type 2 diabetes management: A narrative review. *Diabetes Care, 47*(3), 331–343. https://doi.org/10.2337/dci23-0037
2. Ramos, A. R. (2023). Sleep deprivation, sleep disorders, and chronic disease. *Preventing Chronic Disease, 20.* https://doi.org/10.5888/pcd20.230197
3. Diabetes Care Community. (2024, April 5). *How much sleep do you really need?* https://www.diabetescarecommunity.ca/diet-and-fitness-articles/physical-activity-articles/how-much-sleep-do-you-really-need/
4. Antza, C., Kostopoulos, G., Mostafa, S., Nirantharakumar, K., & Tahrani, A. (2022). The links between sleep duration, obesity and type 2 diabetes mellitus. *Journal of Endocrinology, 252*(2), 125– 141. https://doi.org/10.1530/JOE-21-0155
5. Diabetes UK (2023, August 23). *Sleep and diabetes.* https://www.diabetes.org.uk/guide-to-diabetes/life-with-diabetes/sleep-and-diabetes
6. National Institute of Diabetes and Digestive and Kidney Diseases. (2021). *The impact of poor sleep on type 2 diabetes.* https://www.niddk.nih.gov/health-information/professionals/diabetes-discoveries-practice/the-impact-of-poor-sleep-on-type-2-diabetes#:~:text=Q%3A%20Can%20sleep%20problems%20or,sleep%20all%20promote%20glucose%20intolerance.
7. Rath, L. (2022). Sleep and type 2 diabetes. *WebMD.* https://www.webmd.com/sleep-disorders/sleep-type-2-diabetes
8. Centers for Disease Control and Prevention. (2024, May 15). *New beginnings mini-lesson: Sleep health.* U.S. Department of Health and Human Services. https://www.cdc.gov/diabetes/php/toolkits/new-beginnings-sleep-health.html

Diabetes and Aging

CHAPTER 9

Introduction

As the global population ages, the intersection of diabetes and aging presents unique challenges and opportunities for healthcare management. Chapter 9, "Diabetes and Aging" delves into the complexities older adults face when managing diabetes, offering comprehensive insights and practical strategies to maintain health and independence.

Aging naturally impacts the body's ability to manage blood sugar levels due to changes in metabolism, decreased physical activity, and altered medication requirements. These factors can make diabetes management increasingly challenging for older adults. This chapter explores these changes, addressing common complications such as cardiovascular disease, neuropathy, vision problems, and kidney disease, and offers strategies to mitigate their impact.

Maintaining independence is a crucial aspect of quality of life for older adults with diabetes. Effective medication management, physical activity, and social support are essential components. The chapter highlights the importance of nutrient-dense diets, regular exercise, and using organizational tools like pill organizers to manage multiple medications. It also provides practical tips to prevent falls, enhance heart health, and manage neuropathy symptoms, ensuring older adults can lead active and fulfilling lives.

The chapter further emphasizes the importance of social engagement and mental health. Activities that reduce stress, improve mood, and enhance overall well-being are vital for managing the emotional and psychological burden of diabetes. Strategies for managing cognitive changes, maintaining a healthy weight, and preventing skin problems are discussed, along with tips for handling changes in appetite and bone health.

Dental health, vision care, and hearing loss are also addressed, recognizing that these issues significantly impact the quality of life. The chapter provides actionable advice on maintaining oral hygiene, having regular checkups, and using vision and hearing aids to manage these concerns effectively.

Moreover, the chapter offers guidance on managing blood sugar during illness, adjusting medication needs, and staying physically active despite mobility issues. It underscores the importance of mental and emotional well-being, providing tips for managing depression, loneliness, and stress.

In essence, this chapter is a comprehensive guide for older adults with diabetes and their caregivers, offering practical advice to navigate the complexities of aging while maintaining optimal diabetes management. By following these guidelines, older adults can enhance their quality of life, remain independent, and manage their diabetes effectively.

1. How does aging affect diabetes management?

Aging can make blood sugar control more challenging due to changes in metabolism, physical activity, and medication needs.

2. What are common complications of diabetes in older adults?

Common complications of diabetes in older adults include cardiovascular disease, neuropathy, vision problems such as diabetic retinopathy, and kidney disease. Additionally, older adults with diabetes may also experience an increased risk of infections, poor wound healing, and cognitive decline.

3. How can older adults with diabetes maintain independence?

Older adults with diabetes can maintain independence by effectively managing medications and monitoring blood sugar levels regularly. Staying physically active with appropriate exercises helps maintain mobility and overall health. Additionally, seeking support from healthcare providers, family, or community resources ensures that they have the assistance needed for managing their condition while retaining autonomy.

▲ Figure 9.1
Staying healthy when aging.

4. How can diet needs change for older adults with diabetes?

Diet needs for older adults with diabetes may shift toward focusing on nutrient-dense foods, such as fruits, vegetables, whole grains, and lean proteins, to ensure optimal health. Maintaining a balanced diet that supports blood sugar control, heart health, and digestion is important, and monitoring portion sizes can help prevent overeating and manage blood sugar levels effectively. Additionally, addressing any changes in appetite or chewing/swallowing difficulties may be necessary to ensure proper nutrition.

5. How can physical activity benefit older adults with diabetes?

Exercise improves insulin sensitivity, supports weight management, and enhances overall health.

6. How can older adults with diabetes manage multiple medications?

Use pill organizers, set reminders, and regularly review medications with a healthcare provider.

7. What are some tips for preventing falls in older adults with diabetes?

To prevent falls, ensure a safe home environment by removing tripping hazards, installing grab bars, and using non slip mats. Engaging in balance exercises, such as tai chi or strength training, can improve stability and coordination. Regularly monitoring blood sugar levels is also crucial, as fluctuations in blood sugar can lead to dizziness or weakness, increasing the risk of falls. Additionally, consider using a personal emergency alert system or lifeline to ensure quick access to help in case of a fall.

8. How can older adults with diabetes improve heart health?

Older adults with diabetes can improve heart health by eating a heart-healthy diet rich in fruits, vegetables, whole grains, and healthy fats while limiting saturated fats and sodium. Regular exercise, such as walking or swimming, helps strengthen the heart and improve circulation. It's also essential to manage blood pressure and cholesterol levels through medication, lifestyle changes, and regular checkups with healthcare providers to reduce the risk of cardiovascular complications.

9. How can older adults with diabetes manage neuropathy symptoms?

Older adults with diabetes can manage neuropathy symptoms through medications prescribed by a healthcare provider, such as pain relievers or specific drugs for nerve pain. Physical therapy can help improve mobility and reduce discomfort, while lifestyle changes like maintaining blood sugar control, staying physically active, and wearing comfortable shoes can also alleviate symptoms. Additionally, practices such as regular foot inspections to prevent injury are important for managing neuropathy.

10. What are some ways to protect vision for older adults with diabetes?

To protect vision, older adults with diabetes should have regular eye exams to detect early signs of diabetic retinopathy or other eye conditions. Controlling blood sugar levels and managing blood pressure are crucial in preventing damage to the blood vessels in the eyes. Additionally, adopting a healthy lifestyle, including a balanced diet rich in eye-healthy nutrients like vitamin A and omega-3 fatty acids, can further support vision health.

11. How can older adults with diabetes maintain kidney health?

Older adults with diabetes can maintain kidney health by regularly monitoring kidney function through blood tests, such as checking creatinine levels and glomerular filtration rate (GFR). Managing blood pressure through medication, a healthy diet, and regular physical activity is crucial in preventing kidney damage. Staying hydrated with water and limiting intake of substances that can harm kidney function, like excess salt or alcohol, also helps preserve kidney health.

12. What are some tips for managing stress in older adults with diabetes?

Managing stress in older adults with diabetes can be achieved by practicing relaxation techniques such as deep breathing, meditation, or mindfulness to promote relaxation and reduce anxiety. Staying socially active by maintaining relationships with friends, family, or support groups can provide emotional support and reduce feelings of isolation. Engaging in enjoyable activities, such as hobbies, exercise, or creative pursuits, can also improve mood and help manage stress.

13. How can older adults with diabetes manage cognitive changes?

Stay mentally active with puzzles, reading, or learning new skills; manage blood sugar levels through a balanced diet and regular exercise; and seek support for memory issues through healthcare providers or cognitive training programs.

14. How can older adults with diabetes maintain a healthy weight?

Older adults with diabetes can maintain a healthy weight by following a balanced diet with nutrient-dense foods while being mindful of portion sizes. Staying physically active through regular exercise, such as walking or swimming, supports weight management and overall health. Periodically checking weight, such as once a week or every few weeks, can help track trends and identify any significant changes, with adjustments made as needed. Consulting a healthcare provider for personalized advice is also beneficial.

15. How can older adults with diabetes prevent skin problems?

Keep skin clean and moisturized, monitor for injuries, and manage blood sugar levels.

16. How can older adults with diabetes handle changes in appetite?

Older adults with diabetes can manage changes in appetite by eating smaller, more frequent meals throughout the day to maintain stable blood sugar levels. Focusing on nutrient-dense foods, such as vegetables, lean proteins, and whole grains, ensures they get the necessary nutrients while managing hunger. If appetite changes persist or affect overall health, it's important to consult a healthcare provider for guidance and to rule out any underlying issues.

17. How can older adults with diabetes manage bone health?

Engage in weight-bearing exercises, get adequate calcium and vitamin D, and avoid smoking.

18. How can older adults with diabetes handle dental health?

Maintain good oral hygiene, have regular dental checkups, and manage blood sugar levels.

19. What are some benefits of social activities for older adults with diabetes?

Social activities reduce stress, improve mood, and enhance overall well-being.

20. How can older adults with diabetes ensure they are getting enough sleep?

Maintain a regular sleep schedule, create a comfortable sleep environment, and manage health issues that affect sleep.

21. How can older adults with diabetes manage blood sugar during illness?

Create a personalized sick day plan with your healthcare provider, monitor blood sugar more frequently, stay hydrated, and adjust medications as needed. Follow the specific guidelines for managing diabetes during illness, including when to check blood sugar, when to seek medical help, and how to adjust insulin or other medications.

22. How can older adults with diabetes handle changes in medication needs?

Regularly review medications with a healthcare provider and adjust as needed.

23. What are some tips for staying physically active for older adults with diabetes?

Choose low-impact activities, stay consistent, and include strength and balance exercises.

24. How can older adults with diabetes handle mobility issues?

Use assistive devices if needed, engage in physical therapy, and stay active within limitations.

25. How can older adults with diabetes manage chronic pain?

Use pain management techniques, medications, and lifestyle changes to manage discomfort. Always consult with a healthcare provider to develop a personalized pain management plan.

26. How can older adults with diabetes handle changes in vision?

Older adults with diabetes can manage changes in vision by using vision aids, such as glasses or magnifiers, to improve clarity. Adjusting lighting in the home and workspace can also help reduce eye strain. It's important to have regular eye exams to monitor for diabetic retinopathy and other conditions, ensuring early detection and treatment if needed.

27. How can older adults with diabetes manage hearing loss?

Use hearing aids if prescribed, reduce background noise, and have regular hearing checkups.

28. What are some tips for managing blood sugar levels in older adults?

Monitor levels regularly, follow a balanced diet, and adjust medications as needed, in consultation with a healthcare provider.

29. How can older adults with diabetes improve their immune system?

Older adults with diabetes can strengthen their immune system by getting recommended vaccinations, such as the flu and pneumonia vaccines, to prevent illness. Eating a healthy, balanced diet rich in vitamins and minerals supports immune function, while staying physically active helps improve circulation and overall health. Regular sleep, stress management, and staying hydrated also contribute to a strong immune system.

30. How can older adults with diabetes manage their condition in assisted living facilities?

Communicate needs with caregivers, maintain a routine, and monitor blood sugar levels.

31. How can older adults with diabetes handle dietary restrictions?

Work with a dietitian to create a meal plan that meets nutritional needs and preferences.

32. How can older adults with diabetes ensure they have access to healthy foods?

Older adults with diabetes can ensure access to healthy foods by utilizing grocery delivery services, which can bring fresh produce and other nutritious options directly to their home. Seeking community resources, such as local food banks or senior meal programs, can also provide access to healthy foods. Additionally, planning meals in advance and making use of pre portioned or frozen healthy meal options can help maintain a balanced diet and simplify grocery shopping.

33. What are some tips for managing diabetes during hospital stays?

Communicate with healthcare providers, monitor blood sugar, and follow hospital guidelines.

34. How can older adults with diabetes handle loneliness?

Older adults with diabetes can handle loneliness by staying socially active and engaging in community activities, such as local clubs or volunteer opportunities, to build connections. Joining support groups for people with diabetes or other shared interests can provide emotional support and a sense of belonging. Regular communication with family and friends, whether through phone calls, video chats, or in-person visits, also helps combat feelings of isolation.

35. How can older adults with diabetes manage depression?

Older adults with diabetes can manage depression by seeking therapy, staying socially active, and engaging in activities that bring joy. Connecting with others, whether through support groups or social activities, can help alleviate feelings of loneliness. Consulting a healthcare provider for treatment options, such as medication or therapy, is important for managing both depression and diabetes. Additionally, maintaining a regular exercise routine and focusing on a balanced lifestyle can positively impact mental health.

36. How can older adults with diabetes handle financial challenges?

Older adults with diabetes can manage financial challenges by utilizing community resources, such as food assistance programs or local health services. Seeking financial assistance through government programs, insurance support, or non profit organizations can help ease the burden. It's also important to budget for healthcare needs, including medications, doctor visits, and diabetes management tools, to ensure financial stability while maintaining proper care.

37. How can older adults with diabetes manage their condition during travel?

Plan ahead by packing extra supplies, including medications and snacks, monitor blood sugar levels regularly, and research medical resources at destination.

38. What are some tips for older adults with diabetes to stay hydrated?

Drink water regularly, monitor fluid intake, and avoid excessive caffeine and alcohol.

39. How can older adults with diabetes manage urinary issues?

Older adults with diabetes can manage urinary issues by monitoring fluid intake to avoid dehydration and promote healthy kidney function. Managing blood sugar levels is crucial, as high blood sugar can contribute to urinary problems. It's also important to seek medical advice to address any underlying issues, such as urinary tract infections or bladder problems, and to discuss appropriate treatments or lifestyle adjustments with a healthcare provider.

40. How can older adults with diabetes handle changes in mobility?

Older adults with diabetes can handle changes in mobility by using assistive devices, such as canes or walkers, to improve stability and support. Engaging in physical therapy can help strengthen muscles, improve flexibility, and enhance balance. It's also important to stay as active as possible within physical limits, incorporating gentle exercises like walking or swimming to maintain mobility and overall health. Consulting with healthcare providers for tailored strategies and adaptive equipment may also be beneficial.

41. How can older adults with diabetes handle changes in mental health?

Older adults with diabetes can handle changes in mental health by seeking support from a healthcare provider to address concerns such as depression or anxiety. Staying mentally active through activities like reading, puzzles, or learning new skills can help maintain cognitive function. Managing blood sugar levels is also crucial, as fluctuations can impact mood and mental clarity. Engaging in activities that promote emotional well-being, such as socializing, practicing relaxation techniques, or participating in hobbies, can further support mental health.

42. What are some benefits of staying active for older adults with diabetes?

Physical activity improves insulin sensitivity, supports weight management, and enhances overall health.

43. How can older adults with diabetes manage their condition during emergencies?

Have an emergency plan in place, keep extra supplies (like insulin, glucose tablets, and medications), inform emergency responders of condition, and carry a list of current medications along with important contacts, including healthcare provider information.

44. How can older adults with diabetes ensure they are taking medications correctly?

Use pill organizers, set reminders, and regularly review medications with a healthcare provider.

45. How can older adults with diabetes handle stress?

Practice relaxation techniques, stay socially active, engage in enjoyable activities, and speak with a healthcare provider if stress becomes difficult to manage.

46. How can older adults with diabetes manage their condition during seasonal changes?

During seasonal changes, older adults with diabetes should adjust their routines to accommodate shifts in weather, such as dressing in layers during colder months or staying hydrated in warmer weather. It's important to stay consistent with diabetes management, including monitoring blood sugar levels, adjusting exercise routines, and maintaining a balanced diet. Regular check-ins with healthcare providers can help ensure that any necessary adjustments are made to medication or lifestyle habits.

47. How can older adults with diabetes manage changes in digestion?

Focus on eating fiber-rich foods, stay hydrated, and eat smaller, more frequent meals to support digestion. If issues persist, it's important to consult a healthcare provider to identify any underlying causes or medication side effects.

48. How can older adults with diabetes manage their condition during social gatherings?

Plan ahead by bringing suitable snacks, discuss meal options with hosts if necessary, stay mindful of portion sizes, and monitor blood sugar levels before and after the event to ensure stability.

49. How can older adults with diabetes handle changes in energy levels?

Eat nutrient-rich meals, stay active with regular exercise, monitor blood sugar levels, and ensure adequate rest by maintaining a consistent sleep routine. Consider consulting a healthcare provider if low energy becomes difficult to cope with.

50. What are some ways older adults with diabetes can improve their overall health?

Maintain a balanced diet rich in nutrients, engage in regular physical activity, manage stress through relaxation techniques, prioritize good sleep, and consistently monitor blood sugar levels.

Conclusion

This chapter provides a thorough and insightful examination of the multifaceted challenges faced by older adults managing diabetes. Aging affects diabetes management in various ways, from changes in metabolism and physical activity levels to increased susceptibility to complications like cardiovascular disease, neuropathy, vision problems, and kidney disease. Understanding these impacts is crucial for developing effective strategies to maintain health and independence in older adults with diabetes.

One of the central themes of the chapter is the importance of maintaining independence. Effective medication management, regular physical activity, and a supportive social network are essential. By focusing on nutrient-dense foods, staying physically active, and using organizational tools, older adults can better manage their diabetes and overall health. Practical tips for preventing falls, enhancing heart health, and managing neuropathy symptoms are vital for ensuring that older adults can lead active, fulfilling lives.

The chapter also emphasizes the critical role of social engagement and mental health. Activities that reduce stress, improve mood, and enhance overall well-being are integral to managing the emotional and psychological burdens of diabetes. Strategies for managing cognitive changes, maintaining a healthy weight, and preventing skin problems further contribute to a holistic approach to diabetes care in older adults.

Additionally, the chapter addresses specific health concerns such as dental care, vision, and hearing loss, providing actionable advice on maintaining oral hygiene, having regular checkups, and using vision and hearing aids. These measures significantly impact the quality of life and overall health management.

Managing blood sugar during illness, adjusting medication needs, and staying physically active despite mobility issues are other critical aspects discussed. The importance of mental and emotional well-being is underscored, with tips for managing depression, loneliness, and stress.

Ultimately, this chapter serves as a comprehensive guide for older adults with diabetes and their caregivers. By following the practical advice and strategies outlined, older adults can enhance their quality of life, maintain independence, and effectively manage their diabetes. This holistic approach ensures that older adults can navigate the complexities of aging with confidence and improved well-being.

Bibliography

1. National Council on Aging. (2024, April 5). *What is diabetes? A guide for older adults.* https://www.ncoa.org/article/what-is-diabetes-a-guide-for-older-adults/
2. Healthy Aging Collaborative Online Resources and Education. (n.d.). *Aging and chronic diseases: A profile of Canadian seniors.* https://healthyagingcore.ca/resources/resource-aging-and-chronic-diseases-a-profile-of-canadian-seniors
3. Meneilly, G. S., Knip, A., Miller, D. B., Sherifali, D., Tessier, D., & Zahedi, A. (2018). Diabetes in older people. *Canadian Journal of Diabetes, 42*(S1), S283–S295. https://doi.org/10.1016/j.jcjd.2017.10.021
4. Sinclair, A., Saeedi, P., Kaundal, A., Karuranga, S., Malanda, B., & Williams, R. (2020, April). Diabetes and global ageing among 65-99-year-old adults: Findings from the International Diabetes Federation Diabetes Atlas, 9th edition. *Diabetes research and clinical practice, 162,* 108078. https://doi.org/10.1016/j.diabres.2020.108078
5. Mayo Clinic. (2018, March 1). *Life with diabetes: What happens as we age?* https://www.mayoclinichealthsystem.org/hometown-health/speaking-of-health/life-with-diabetes-what-happens-as-we-age
6. National Institute on Aging. (2024, April 10). *Diabetes in older people.* https://www.nia.nih.gov/health/diabetes/diabetes-older-people
7. WebMD. (2024, June 15). *How age relates to type 2 diabetes.* https://www.webmd.com/diabetes/diabetes-link-age

Diabetes and Mental Health

CHAPTER 10

Introduction

Mental health is a crucial aspect of diabetes management that profoundly impacts overall well-being. In this chapter, we delve into the intricate relationship between diabetes and mental health, exploring various challenges individuals may face and effective strategies for coping and support.

Living with diabetes involves more than just managing blood sugar levels; it encompasses navigating emotional hurdles such as stress, anxiety, depression, and diabetes-specific distress. The demands of daily self-care, including monitoring blood glucose, medication adherence, and dietary considerations, can create a significant emotional burden. Understanding these challenges is essential for individuals with diabetes and their support networks to effectively address mental health concerns.

Recognizing the signs of mental health issues, such as depression and anxiety, is crucial. From persistent sadness to changes in sleep patterns and appetite, these symptoms can impact diabetes management and overall quality of life. Equipping oneself with coping strategies, accessing professional support like therapy or counseling, and integrating stress-reduction techniques are vital steps toward managing these challenges effectively.

Moreover, fostering a supportive environment plays a pivotal role in mental health maintenance. Social support from family, friends, and peers can provide emotional comfort, practical assistance, and reduce feelings of isolation. Building a strong support network and engaging in peer support groups or online communities can offer valuable insights and solidarity in navigating the emotional complexities of diabetes.

Throughout this chapter, we will explore practical guidance on enhancing emotional well-being, managing stress, improving self-esteem, and cultivating resilience. By addressing the intersection of diabetes and mental health holistically, individuals with diabetes can strive toward not only managing their physical health but also nurturing their mental and emotional wellness for a fulfilling life.

1. How does diabetes affect mental health?

Managing diabetes can lead to stress, anxiety, and depression due to the constant attention required for blood sugar control.

2. What is diabetes distress and how can it be managed?

Diabetes distress refers to the emotional burden of managing diabetes. It can be managed through support groups, therapy, and stress-reduction techniques.

3. How can people with diabetes recognize signs of depression?

Signs include persistent sadness, loss of interest in activities, changes in sleep and appetite, and feelings of hopelessness.

4. What are some coping strategies for anxiety and stress in people with diabetes?

Coping strategies include deep breathing exercises, mindfulness, physical activity, and seeking professional help if needed.

▲ **Figure 10.1**
Deep breathing.

5. How can cognitive behavioral therapy (CBT) help people with diabetes?

CBT can help by changing negative thought patterns, improving problem-solving skills, and reducing stress.

6. What is the relationship between blood sugar levels and mood?

Fluctuating blood sugar levels can cause mood swings, irritability, and anxiety.

7. How can people with diabetes improve their emotional well-being?

Regular physical activity, a balanced diet, adequate sleep, and engaging in enjoyable activities can improve emotional well-being.

8. What role does social support play in managing diabetes and mental health?

Social support provides emotional comfort, practical help, and can reduce feelings of isolation and stress.

9. How can people with diabetes handle feelings of burnout?

Take breaks, seek support, focus on self-care, and consult healthcare providers for guidance.

10. How can mindfulness and meditation benefit people with diabetes?

These practices reduce stress, improve focus, and promote relaxation, which can help in managing diabetes.

11. How can people with diabetes build a strong support network?

Connect with family and friends, join support groups, and engage with online communities.

12. What are some tips for improving self-esteem in people with diabetes?

Focus on achievements, set realistic goals, and practice self-compassion.

13. How can people with diabetes handle negative emotions?

Acknowledge emotions, practice stress-reduction techniques, and seek support if needed.

14. How can people with diabetes develop resilience?

Build coping skills, maintain a positive outlook, and seek support during challenging times.

15. How can people with diabetes balance their mental and physical health?

Maintain a healthy lifestyle, monitor blood sugar levels, and prioritize mental health care.

16. What are some signs that people with diabetes need professional mental health support?

Persistent sadness, overwhelming anxiety, difficulty managing diabetes, and thoughts of self-harm.

17. How can people with diabetes manage emotional eating?

Identify triggers, practice mindful eating, and seek professional help if needed.

18. How can people with diabetes stay motivated in their self-care routine?

Set small, achievable goals, reward progress, and seek support from healthcare providers and loved ones.

19. How can people with diabetes handle feelings of guilt or blame?

Practice self-compassion, focus on positive changes, and seek support if needed.

20. How can people with diabetes improve their relationship with food?

Practice mindful eating, avoid restrictive diets, and seek guidance from a dietitian.

21. How can people with diabetes manage fear of hypoglycemia?

Monitor blood sugar levels regularly, carry glucose tablets, and develop a plan for managing lows.

22. What are some relaxation techniques for people with diabetes?

Techniques include deep breathing exercises, progressive muscle relaxation, and guided imagery.

23. How can people with diabetes handle social stigma?

Educate others about diabetes, build a supportive network, and practice self-advocacy.

24. How can gratitude practices benefit people with diabetes?

Gratitude practices can improve mood, reduce stress, and enhance overall well-being.

25. How can people with diabetes maintain a positive outlook?

Focus on strengths, set realistic goals, and engage in activities that bring joy.

26. How can people with diabetes manage anger and frustration?

Acknowledge feelings, practice stress-reduction techniques, and seek professional help if needed.

27. How can people with diabetes improve their focus and concentration?

Manage blood sugar levels, get adequate sleep, and engage in cognitive exercises.

28. How can people with diabetes practice self-compassion?

Be kind to yourself, acknowledge your efforts, and avoid self-criticism.

29. How can people with diabetes handle feelings of isolation?

Stay connected with others, join support groups, and seek social activities.

30. How can people with diabetes improve their coping skills?

Practice stress management techniques, seek support, and focus on problem-solving.

31. How can people with diabetes handle changes in routine?

Stay flexible, plan ahead, and seek support if needed.

32. How can people with diabetes manage work-related stress?

Practice time management, take breaks, and seek support from colleagues and supervisors.

33. How can people with diabetes improve their quality of life?

Maintain a healthy lifestyle, seek social support, and focus on overall well-being.

34. How can people with diabetes handle disappointment?

Focus on positive outcomes, seek support, and practice resilience.

35. How can people with diabetes handle financial stress?

Seek financial assistance, create a budget, and explore community resources.

36. How can people with diabetes improve their mental health through hobbies?

Engage in enjoyable activities, explore new interests, and connect with others.

37. How can people with diabetes handle relationship stress?

Communicate openly, seek support, and engage in relationship-building activities.

38. How can people with diabetes improve their self-awareness?

Practice mindfulness, keep a journal, and seek feedback from others.

39. How can people with diabetes manage stress during holidays and celebrations?

Plan ahead, set boundaries, and focus on self-care.

40. How can people with diabetes handle major life changes?

Stay flexible, seek support, and focus on problem-solving.

41. How can people with diabetes improve their mental health through creative activities?

Engage in art, music, or writing to express emotions and reduce stress.

42. How can people with diabetes improve their mental health through nature?

Spend time outdoors, engage in physical activities, and practice mindfulness in nature.

43. How can people with diabetes handle unexpected challenges?

Stay flexible, seek support, and focus on problem-solving.

44. How can people with diabetes manage their mental health during caregiving responsibilities?

Seek support, prioritize self-care, and engage in stress-reduction techniques.

45. How can people with diabetes improve their mental health through volunteering?

Volunteering provides a sense of purpose, reduces stress, and connects with others.

46. How can people with diabetes manage their mental health during times of uncertainty?

Stay informed, seek support, and focus on self-care.

47. How can people with diabetes celebrate their mental health progress?

Recognize achievements, reward yourself, and reflect on positive changes.

Conclusion

In conclusion, the interplay between diabetes and mental health is a critical aspect of overall well-being that demands attention and care. This chapter has highlighted the multifaceted challenges faced by individuals with diabetes, emphasizing the importance of recognizing and addressing mental health concerns alongside physical health management.

Living with diabetes requires constant vigilance and self-management, which can lead to significant stress, anxiety, and even depression. Understanding the signs of mental health issues and implementing effective coping strategies is essential for maintaining both emotional and physical health. Recognizing symptoms such as persistent sadness, loss of interest in activities, and changes in sleep and appetite can help individuals seek timely intervention and support.

Coping strategies, including mindfulness, physical activity, cognitive behavioral therapy (CBT), and relaxation techniques, play a pivotal role in managing diabetes-related stress and anxiety. These approaches not only improve emotional well-being but also enhance overall diabetes management by fostering a positive and resilient mindset.

Social support emerges as a cornerstone of effective diabetes and mental health management. The encouragement and understanding of family, friends, and peers can significantly alleviate the emotional burden of diabetes. Building a robust support network and participating in support groups or online communities provide individuals with a sense of belonging and shared experience, which can be incredibly empowering.

Moreover, integrating mental health care into diabetes management plans is crucial. Therapy, particularly CBT, offers valuable tools for changing negative thought patterns, improving problem-solving skills, and reducing stress. Professional support helps individuals navigate the emotional complexities of diabetes, promoting a balanced approach to health and wellness.

In sum, addressing mental health is indispensable for those living with diabetes. By fostering a holistic approach that encompasses physical, emotional, and social well-being, individuals with diabetes can achieve a higher quality of life. Embracing strategies for emotional resilience, seeking support, and prioritizing self-care are essential steps toward managing the intricate relationship between diabetes and mental health.

Bibliography

1. American Diabetes Association. (n.d.). *Mental health.* https://diabetes.org/health-wellness/mental-health
2. Centers for Disease Control and Prevention. (2024, May 15). *Diabetes and mental health.* https://www.cdc.gov/diabetes/living-with/mental-health.html
3. Diabetes Canada. (n.d.). *How diabetes can affect your mental health.* https://www.diabetes.ca/archive/managing-my-diabetes---archive/preventing-complications/mental-health
4. Garrett, C., & Doherty, A. (2014). Diabetes and mental health. *Clinical medicine (London, England), 14*(6), 669–672. https://doi.org/10.7861/clinmedicine.14-6-669
5. Robinson, D. J., Hanson, K., Jain, A. B., Kichler, J. C., Mehta, G., Melamed, O. C., Vallis, M., Bajaj, H. S., Barnes, T., Gilbert, J., Honshorst, K., Houlden, R., Kim, J., Lewis, J., MacDonald, B., MacKay, D., Mansell, K., Rabi, D., Sherifali, D., & Senior, P. (2023). *Diabetes and mental health. Canadian Journal of Diabetes, 47*(4), 308–344. https://doi.org/10.1016/j.jcjd.2023.04.009
6. WebMD. (2024). *What to know about diabetes and depression.* https://www.webmd.com/diabetes/what-to-know-about-depression-and-diabetes
7. Mental Health America. *Diabetes and mental health.* https://www.mhanational.org/diabetes-and-mental-health
8. Kang W. (2022). Investigating the association between diabetes and mental health: A train-and-test approach. *Frontiers in psychiatry, 13,* 1044714. https://doi.org/10.3389/fpsyt.2022.1044714

Diabetes and Family Life

CHAPTER 11

Introduction

In the realm of diabetes management, the role of family support is not just crucial but often transformative. Families play an integral part in creating a nurturing environment that supports the physical, emotional, and practical needs of individuals living with diabetes. This chapter explores a myriad of ways families can actively engage in and enhance the diabetes management journey, fostering a sense of teamwork and empowerment within the family unit.

Understanding diabetes starts with education, and family members who equip themselves with knowledge about the condition can provide invaluable support. From learning about blood sugar monitoring to understanding medication regimens, informed family members can assist in practical aspects of daily management. Moreover, emotional support plays a pivotal role. Families can offer encouragement, empathy, and a listening ear, helping to alleviate the emotional burden that often accompanies diabetes.

Creating a supportive environment involves promoting healthy lifestyle habits across the family spectrum. Whether it's through shared meal planning, encouraging physical activity, or navigating special occasions with mindful choices, families can collectively foster a diabetes-friendly lifestyle. Open communication is key, enabling discussions about challenges, adjustments, and shared goals related to diabetes management.

For families with children diagnosed with diabetes, the dynamics shift to include age-appropriate education and tailored support strategies. From school-related diabetes management to navigating the emotional impact on siblings, understanding and proactive involvement are essential. Similarly, supporting adults with diabetes may involve navigating career challenges, financial considerations, and lifestyle adjustments.

Throughout this chapter, we explore practical tips, emotional strategies, and proactive approaches that families can integrate into their daily lives to support their loved ones effectively. By cultivating a supportive atmosphere that values education, communication, and holistic health, families can empower individuals with diabetes to thrive and celebrate successes together. This journey not only strengthens familial bonds but also enhances overall well-being, emphasizing that diabetes management is a collaborative effort where every family member plays a crucial role.

1. How can family members support someone with diabetes?

Educate themselves about diabetes, participate in healthy activities, provide emotional support, and assist with diabetes management tasks.

They can create a supportive environment for their loved ones with diabetes by promoting healthy eating, encouraging physical activity, and maintaining open communication.

For parents wanting to support their child with diabetes management, they can educate the child about diabetes, monitor blood sugar levels, and collaborate with healthcare providers.

Couples managing their diabetes together should communicate openly, support each other's health goals, and work together on meal planning and physical activity.

2. How can families handle the emotional impact of a diabetes diagnosis?

Seek support, communicate openly, and focus on coping strategies.

3. How can families create a diabetes-friendly meal plan?

Plan balanced meals, involve everyone in meal preparation, and focus on healthy eating habits.

▲ Figure 11.1
Cooking with family.

4. How can families support physical activity for someone with diabetes?

Participate in activities together, encourage regular exercise, and find enjoyable ways to stay active.

5. How can families handle diabetes-related emergencies?

Have a plan in place, keep emergency supplies accessible, and educate all family members about what to do.

6. How can families help with medication management for someone with diabetes?

Set reminders, assist with organizing medications, and ensure prescriptions are refilled on time.

7. How can families support diabetes management during holidays and special occasions?

Plan ahead, make healthy food choices, and balance indulgence with diabetes management.

8. How can families handle the stress of diabetes management?

Practice stress- reduction techniques, seek support, and maintain open communication.

9. How can families support a teenager with diabetes?

Educate the teenager about diabetes, encourage independence, and provide emotional support.

10. How can families handle school-related diabetes management?

Communicate with school staff, create a diabetes management plan, and provide necessary supplies.

11. How can families support diabetes management in older adults?

Assist with medication management, monitor blood sugar levels, and provide emotional support.

12. How can families handle diabetes-related conflicts?

Communicate openly, practice empathy, and seek solutions together.

13. How can families help with blood sugar monitoring for someone with diabetes?

Set reminders, assist with monitoring, and keep track of blood sugar levels.

14. How can families support a loved one with diabetes during illness?

Monitor blood sugar levels closely, ensure proper hydration, and follow sick day guidelines.

15. How can families help with dietary restrictions for someone with diabetes?

Plan meals together, focus on healthy food choices, and find alternatives for restricted foods.

16. How can families support diabetes management during travel?

Plan ahead, pack necessary supplies, and monitor blood sugar levels.

17. How can families handle diabetes-related financial challenges?

Budget for diabetes supplies, seek financial assistance, and explore community resources.

18. How can families support mental health for someone with diabetes?

Provide emotional support, encourage healthy coping strategies, and seek professional help if needed.

19. How can families handle changes in diabetes management needs?

Stay informed, communicate openly, and adapt to new routines.

20. How can families support diabetes education and awareness?

Educate themselves, participate in awareness activities, and share information with others.

21. How can families support a loved one with diabetes in social situations?

Plan ahead, make healthy choices, and provide support during social events.

22. How can families handle the impact of diabetes on family dynamics?

Communicate openly, seek support, and focus on problem-solving together.

23. How can families support diabetes management during pregnancy?

Monitor blood sugar levels closely, follow healthcare provider guidelines, and provide emotional support.

24. How can families help with foot care for someone with diabetes?

Encourage regular foot inspections, assist with proper footwear, and seek medical advice for foot issues.

25. How can families support diabetes management in the workplace?

Encourage healthy habits, provide emotional support, and help with planning and organization.

26. How can families handle diabetes-related fatigue?

Encourage rest, manage stress, and focus on healthy habits.

27. How can families support diabetes management during weight loss efforts?

Plan balanced meals, encourage physical activity, and provide emotional support.

28. How can families help with vision care for someone with diabetes?

Encourage regular eye exams, assist with vision aids, and monitor blood sugar levels.

29. How can families support diabetes management during medication changes?

Stay informed, assist with organizing medications, and monitor for side effects.

30. How can families handle the impact of diabetes on relationships?

Communicate openly, practice empathy, and seek solutions together.

31. How can families support diabetes management during caregiving responsibilities?

Share caregiving tasks, seek support, and focus on self-care.

32. How can families help with managing diabetes technology?

Stay informed about new technologies, assist with device management, and provide support.

33. How can families support diabetes management during physical therapy?

Encourage participation, assist with exercises, and monitor blood sugar levels.

34. How can families handle the emotional impact of diabetes complications?

Seek support, communicate openly, and focus on coping strategies.

35. How can families support diabetes management during changes in routine?

Stay flexible, plan ahead, and adapt to new routines.

36. How can families help with diabetes management in assisted living facilities?

Communicate with caregivers, ensure proper diabetes care, and provide emotional support.

37. How can families support diabetes management during hospitalization?

Communicate with healthcare providers, monitor blood sugar levels, and follow hospital guidelines.

38. How can families support diabetes management during sports and activities?

Plan ahead, monitor blood sugar levels, and provide support during activities.

39. How can families help with managing diabetes supplies?

Keep track of supplies, set reminders for refills, and ensure proper storage.

40. How can families support diabetes management during lifestyle changes?

Communicate openly, adapt to new routines, and focus on healthy habits.

41. How can families handle diabetes-related conflicts with healthcare providers?

Communicate openly, seek second opinions if needed, and focus on finding solutions.

42. How can families celebrate successes in diabetes management?

Recognize achievements, reward progress, and focus on positive changes.

Conclusion

Family involvement in diabetes management is pivotal for creating a supportive and nurturing environment that enhances overall well-being and quality of life. Throughout this chapter, we have explored numerous ways in which families can actively engage and support their loved ones with diabetes, addressing both the practical and emotional aspects of care.

Education emerges as a cornerstone for effective support. When family members understand the intricacies of diabetes, from monitoring blood sugar levels to administering medications, they can provide valuable assistance in daily management tasks. This knowledge also enables families to better navigate challenges and make informed decisions alongside healthcare providers.

Emotional support plays an equally vital role. Diabetes can bring about significant emotional challenges, including stress, anxiety, and feelings of isolation. By offering encouragement, empathy, and a nonjudgmental space for open communication, families can help alleviate these emotional burdens and strengthen bonds within the family unit.

Creating a diabetes-friendly environment involves fostering healthy habits across all family members. Whether through collaborative meal planning, promoting regular physical activity, or navigating social situations with mindfulness, families can collectively embrace a lifestyle that supports diabetes management. This holistic approach not only benefits the individual with diabetes but promotes overall family health and well-being.

Flexibility and adaptability are also key themes. Diabetes management requires ongoing adjustments in response to changes in routines, health status, and treatment plans. Families that remain flexible, open to learning, and proactive in problem-solving can effectively navigate these fluctuations and support their loved ones through various life stages and challenges.

Celebrating successes is essential in fostering motivation and resilience. Recognizing milestones, progress in management goals, and the everyday victories helps individuals with diabetes feel supported and encouraged on their journey. It reinforces a positive mindset and reinforces the importance of ongoing efforts in managing diabetes effectively.

Ultimately, the journey of diabetes management is one that involves the entire family. By embracing education, communication, healthy habits, and emotional support, families can create a nurturing environment where individuals with diabetes not only manage their condition but thrive. This collaborative approach not only strengthens familial bonds but also empowers individuals with diabetes to lead fulfilling and healthy lives.

Bibliography

Bibliography

1. American Diabetes Association. (n.d.). *For caregivers.* https://diabetes.org/tools-resources/for-caregivers
2. Centers for Disease Control and Prevention. (2024, May 15). *Helping friends and family with diabetes.* https://www.cdc.gov/diabetes/caring/index.html
3. Diabetes Canada. (2021, February 11). *10 ways to support your loved one with diabetes.* https://www.diabetes.ca/about-diabetes/stories/10-ways-to-support-your-loved-one-with-diabetes
4. Baig, A. A., Benitez, A., Quinn, M. T., & Burnet, D. L. (2015, August 6). Family interventions to improve diabetes outcomes for adults. *Annals of the New York Academy of Sciences, 1353*(1), 89–112. https://doi.org/10.1111/nyas.12844
5. Miller, T. A., & Dimatteo, M. R. (2013, July 29). Importance of family/social support and impact on adherence to diabetic therapy. *Diabetes, metabolic syndrome and obesity : Targets and therapy, 6,* 421–426. https://doi.org/10.2147/DMSO.S36368
6. Kocurek, B. (2021, June 11). *6 ways to support your loved one with diabetes.* Baylor Scott & White Health. https://www.bswhealth.com/blog/6-ways-to-support-your-loved-one-with-diabetes
7. American Association for Marriage and Family Therapy. (n.d.). *Diabetes.* https://www.aamft.org/AAMFT/Consumer_Updates/Diabetes.aspx
8. Hill-Briggs, F., Adler, N. E., Berkowitz, S. A., Chin, M. H., Gary-Webb, T. L., Navas-Acien, A., Thornton, P. L., & Haire-Joshu, D. (2021). Social determinants of health and diabetes: A scientific review. *Diabetes Care, 44*(1), 258–279. https://doi.org/10.2337/dci20-0053

Diabetes and Technology

CHAPTER 12

Introduction

Technology plays an increasingly vital role in the management of diabetes, offering innovative solutions that enhance the quality of life for millions worldwide. This chapter delves into how technology can revolutionize diabetes care, providing tools and resources that empower individuals to take control of their health. From continuous glucose monitors (CGMs) that provide real-time data on blood sugar levels to smartphone apps that track medication and diet, the integration of technology in diabetes management is transforming traditional approaches and making it easier to maintain optimal health.

Continuous glucose monitors (CGMs) and insulin pumps are among the most significant advancements, offering continuous blood sugar monitoring and precise insulin delivery. These devices not only improve blood sugar control but also provide valuable data that can be shared with healthcare providers for more informed decision-making. Additionally, smartphone apps offer a range of features, including blood sugar tracking, carb counting, and medication reminders, all designed to streamline diabetes management and provide educational resources at users' fingertips.

Telemedicine has emerged as a crucial tool, especially in the wake of global health challenges, enabling remote consultations with healthcare providers. This accessibility ensures that individuals with diabetes can receive timely care without the need for frequent in-person visits. Wearable technology, such as smartwatches, further supports diabetes management by tracking physical activity, heart rate, and integrating with diabetes apps for comprehensive health monitoring.

As technology continues to evolve, new trends such as artificial intelligence (AI) and virtual reality (VR) are poised to offer even more sophisticated tools for diabetes management and education. AI can analyze blood sugar data and provide personalized recommendations, while VR can offer immersive educational experiences to enhance learning.

However, with the benefits come challenges, including the cost of devices, technical issues, and data security concerns. This chapter addresses these challenges and offers practical tips for choosing the right technology, maintaining devices, and protecting personal health data.

By exploring the various technological tools and resources available, individuals with diabetes can find solutions tailored to their needs, ensuring better management of their condition and improving their overall quality of life.

1. How can technology help manage diabetes?

Technology can assist with monitoring blood sugar levels, tracking medication, and providing data for better diabetes management.

2. What are continuous glucose monitors (CGMs) and how do they work?

CGMs are devices that measure blood sugar levels continuously throughout the day and night, providing real-time data.

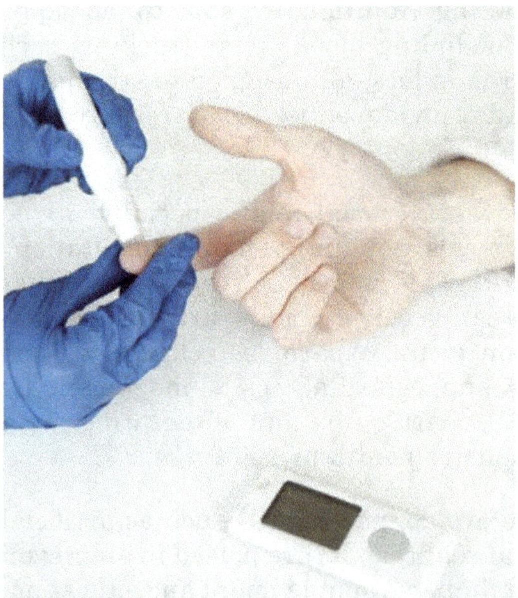

▲ Figure 12.1
Continuous glucose monitor.

3. What are insulin pumps and how do they benefit people with diabetes?

Insulin pumps deliver a continuous dose of insulin and can be adjusted for meals and physical activity, providing more precise insulin management.

4. How can smartphone apps help with diabetes management?

Apps can track blood sugar levels, monitor food intake, remind users to take medication, and provide educational resources.

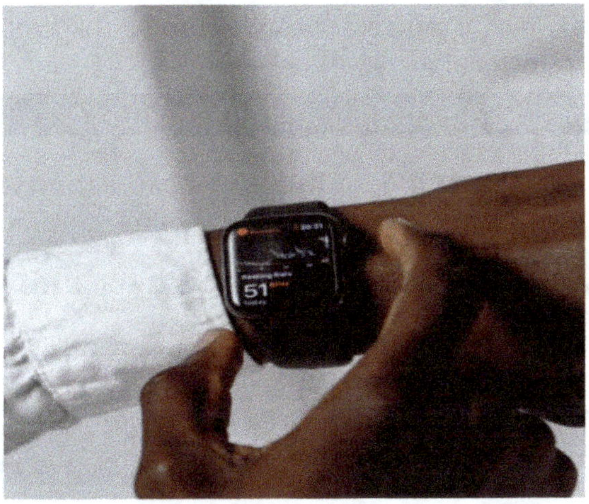

▲ Figure 12.2
Health apps.

5. What are some benefits of using a diabetes management platform?

These platforms integrate data from various devices, provide insights into blood sugar trends, and facilitate communication with healthcare providers.

6. How can telemedicine benefit people with diabetes?

Telemedicine allows for remote consultations with healthcare providers, making it easier to manage diabetes from home.

7. What are some features to look for in a diabetes app?

Look for features like blood sugar tracking, carb counting, medication reminders, and integration with other diabetes devices.

8. How can people with diabetes protect their health data?

Use secure passwords, enable two-factor authentication, and ensure that apps and devices comply with privacy regulations.

9. How can technology help with diabetes education?

Online courses, webinars, and educational apps provide information and resources for better diabetes management.

10. How can people with diabetes stay updated on new technologies?

Follow diabetes organizations, attend conferences, and subscribe to health technology newsletters.

11. How can wearable technology help manage diabetes?

Wearables like smartwatches can monitor physical activity, track heart rate, and integrate with diabetes management apps.

12. What are some challenges of using diabetes technology?

Challenges include device cost, technical issues, data accuracy, and the need for regular maintenance.

13. How can people with diabetes choose the right technology for their needs?

Consult with healthcare providers, research different options, and consider factors like cost, ease of use, and compatibility.

14. What are some benefits of using cloud-based diabetes management systems?

These systems store data securely, allow for easy access from multiple devices, and facilitate data sharing with healthcare providers.

15. How can people with diabetes troubleshoot technology issues?

Follow manufacturer guidelines, contact customer support, and consult with healthcare providers if needed.

16. How can people with diabetes integrate multiple technologies for better management?

Use compatible devices and apps, integrate data into a single platform, and regularly review data with healthcare providers.

17. What are some future trends in diabetes technology?

Trends include advancements in artificial intelligence, new wearable devices, and improved integration of diabetes management tools.

18. How can virtual reality (VR) be used in diabetes education?

VR can provide immersive educational experiences, simulate diabetes management scenarios, and enhance learning.

19. How can people with diabetes use social media for support?

Join online communities, follow diabetes influencers, and participate in discussions to share experiences and gain support.

20. What are some benefits of using a smart insulin pen?

Smart insulin pens track insulin doses, remind users to take their medication, and provide data for better diabetes management.

21. How can people with diabetes use artificial intelligence (AI) for management?

AI can analyze blood sugar data, predict trends, and provide personalized recommendations for better management.

22. What are some tips for maintaining diabetes technology?

Regularly update software, follow manufacturer maintenance guidelines, and store devices properly.

23. How can people with diabetes use technology to monitor diet and nutrition?

Use apps to track food intake, calculate carb counts, and monitor nutritional information.

24. How can technology help with diabetes-related stress?

Apps and devices can provide relaxation techniques, stress tracking, and resources for mental health support.

25. How can people with diabetes use technology to stay active?

Use fitness trackers, activity apps, and online workout programs to monitor and encourage physical activity.

26. How can people with diabetes use technology for medication management?

Use apps to set reminders, track medication doses, and monitor prescription refills.

27. How can smart home devices help with diabetes management?

Smart home devices can remind users to take medication, monitor physical activity, and provide access to health data.

28. How can people with diabetes use technology to improve sleep?

Use sleep tracking devices and apps to monitor sleep patterns, improve sleep hygiene, and address sleep issues.

29. How can people with diabetes use technology to monitor their heart health?

Use wearable devices to track heart rate, monitor blood pressure, and provide data for better cardiovascular health management.

30. How can technology help with managing diabetes-related complications?

Devices can monitor for signs of complications, provide early warnings, and facilitate communication with healthcare providers.

31. How can people with diabetes use technology for goal setting?

Use apps and devices to set and track health goals, monitor progress, and adjust plans as needed.

32. How can people with diabetes use technology to manage their care team?

Use platforms that facilitate communication, share data, and coordinate care with multiple healthcare providers.

33. How can people with diabetes use technology to stay organized?

Use apps and devices to track appointments, manage supplies, and keep all diabetes-related information in one place.

34. How can technology help with diabetes research?

Data from devices and apps can contribute to research studies, improving understanding and treatment of diabetes.

35. How can people with diabetes use technology to stay motivated?

Use apps that provide reminders, track progress, and offer rewards for achieving health goals.

36. How can people with diabetes use technology for self-monitoring?

Regularly review data from devices and apps, identify patterns, and adjust management plans as needed.

37. How can people with diabetes use technology to improve their quality of life?

Technology can simplify diabetes management, reduce stress, and provide resources for better overall health.

38. How can people with diabetes use technology to access healthcare resources?

Use apps and online platforms to find healthcare providers, schedule appointments, and access educational materials.

39. How can people with diabetes use technology to track symptoms?

Use apps to log symptoms, monitor changes, and share information with healthcare providers.

40. How can technology help with diabetes prevention?

Technology can provide education, track health metrics, and encourage healthy lifestyle choices to prevent diabetes.

41. How can people with diabetes use technology to stay informed about their condition?

Follow reputable sources, subscribe to newsletters, and participate in online courses for the latest information.

42. How can people with diabetes use technology to monitor their weight?

Use smart scales and apps to track weight changes, set goals, and monitor progress.

43. How can people with diabetes use technology to improve their hydration?

Use apps to track water intake, set reminders, and monitor hydration levels.

44. How can people with diabetes use technology to manage their appointments?

Use scheduling apps and reminders to keep track of healthcare appointments and follow-up visits.

45. How can people with diabetes use technology to track their progress?

Regularly review data from devices and apps, identify trends, and celebrate achievements.

46. How can people with diabetes use technology to improve their diet?

Use apps to track food intake, find healthy recipes, and monitor nutritional information.

47. How can people with diabetes use technology to manage their condition during travel?

Use apps to plan meals, monitor blood sugar levels, and access healthcare resources while traveling.

48. How can people with diabetes use technology to improve their overall health?

Integrate various health apps and devices to monitor all aspects of health, from blood sugar levels to physical activity and sleep.

49. How do continuous glucose monitors (CGMs) improve diabetes management?

Continuous glucose monitors (CGMs) play a transformative role in diabetes management by providing real-time glucose readings and trends throughout the day and night. Unlike traditional glucose meters that require finger pricks to measure blood sugar, CGMs continuously track glucose levels through a small sensor inserted under the skin. The benefits of CGMs include:

- **Real-time data:** CGMs allow users to see current glucose levels and track trends, offering insights into how food, exercise, medication, and stress impact glucose.
- **Trend analysis:** By displaying glucose trends and rate of change, CGMs help users understand fluctuations in blood sugar, allowing them to take proactive steps if levels are rising or falling.
- **Alerts for hypo- and hyperglycemia:** CGMs provide real-time alerts when glucose levels approach dangerously high or low ranges, which can be lifesaving, especially for people with hypoglycemia unawareness.
- **Improved time in range (TIR):** CGM users often achieve a higher TIR, meaning they maintain glucose levels within their target range more consistently, leading to better overall glycemic control and reduced risk of complications.
- **Data sharing and remote monitoring:** Many CGMs allow data to be shared with caregivers, family members, or healthcare providers, enabling a supportive network and timely interventions.

By reducing the need for constant finger pricks, CGMs improve comfort, convenience, and empower users to make informed decisions about their diabetes management.

50. What are the key features to consider when choosing a CGM or insulin pump?

Selecting the right CGM or insulin pump depends on individual needs, lifestyle, and medical requirements. Key features to consider include:

- **Accuracy and reliability:** The accuracy of CGM readings is paramount. Certain models, like the Dexcom G6 and Freestyle Libre, are known for high accuracy without requiring finger-stick calibration.
- **Integration with insulin pumps:** Some CGMs integrate directly with insulin pumps, allowing for automated insulin delivery adjustments. This setup, often called a closed-loop system or artificial pancreas, reduces manual insulin calculations.
- **Alert customization:** Users should look for CGMs that offer customizable alerts for high and low glucose levels to suit their lifestyle and risk level. This is especially important for people with hypoglycemia unawareness.
- **Data sharing options:** CGMs with remote sharing capabilities allow caregivers or family members to receive alerts and monitor glucose trends, which can be crucial for children, elderly individuals, or those who live alone.
- **Battery life and durability:** Battery life and sensor wear time can vary significantly. Longer wear times reduce sensor changes, but durability is crucial, especially for active users.
- **User interface and app compatibility:** The user interface on associated apps or receivers should be intuitive, with data easily accessible. Compatibility with smartphones and wearable devices is a plus.
- **Cost and insurance coverage:** The cost of devices and supplies, as well as insurance coverage, play a major role in device selection. Some CGMs and insulin pumps can be costly, so insurance benefits should be verified beforehand.

Each CGM and insulin pump has unique features, so users should weigh their options to find a device that fits seamlessly with their daily routine and diabetes management goals.

51 . How can CGM data be used to adjust insulin dosing and dietary choices?

CGM data provides real-time feedback on how diet, exercise, and insulin doses affect glucose levels, making it a powerful tool for fine-tuning diabetes management:

- **Bolus calculations for meals:** By tracking how blood glucose responds to meals, users can adjust their insulin-to-carb ratios more accurately, leading to more precise bolus insulin dosing.
- **Pattern recognition:** Over time, CGM data reveals patterns in glucose levels, such as the dawn phenomenon (morning glucose rises) or nocturnal hypoglycemia. Users can work with healthcare providers to adjust basal insulin or modify bedtime snacks accordingly.
- **Carbohydrate timing and portion control:** Observing post meal glucose spikes helps users make smarter carbohydrate choices. For instance, users may adjust portion sizes, choose low glycemic index (GI) foods, or distribute carbs more evenly throughout the day.
- **Exercise adjustments:** CGMs help users determine how different types of physical activity impact glucose levels. This insight allows for adjustments in carb intake, exercise timing, or basal insulin settings.
- **Preventing hypo- and hyperglycemia:** Real-time glucose readings help users adjust insulin dosing to prevent low or high glucose levels. For example, if glucose is trending low, users can proactively consume carbs to avoid hypoglycemia.

Using CGM data to inform dietary and insulin decisions enables individuals to personalize their management plan, achieve better glycemic control, and ultimately reduce the risk of complications.

52. What should be considered when integrating mobile health apps for diabetes tracking?

Mobile health apps can greatly assist in tracking and managing diabetes, but a few considerations ensure the apps are beneficial:

- **Data integration:** Apps that integrate with other diabetes devices, such as CGMs and insulin pumps, streamline tracking and enable easy data analysis.
- **User-friendly interface:** Look for apps with an intuitive interface, customizable settings, and accessible information on glucose readings, medication reminders, meal tracking, and activity logs.
- **Data security:** Since health apps store sensitive data, check that the app complies with privacy regulations like HIPAA. Security measures like encryption are important for data protection.
- **Educational resources:** Some apps offer diabetes education, recipes, or exercise tips. This extra content can provide a valuable learning resource.
- **Customer support:** Responsive customer support is important if app issues arise, or technical assistance is needed.
- **Cost and subscription model:** Some apps are free, while others require a subscription or in-app purchases. Assess the app's cost versus the benefit it provides.
- **Sharing capabilities:** Apps that allow data sharing with healthcare providers and family can enhance collaborative management and receive timely guidance.

Choosing the right app ensures that it aligns well with the user's management style and daily routine and provides seamless support for diabetes tracking.

53 . What are some best practices for using technology effectively in diabetes management?

To maximize the benefits of diabetes management technology, follow these best practices:

- **Regularly calibrate devices:** If required, calibrate CGMs and other devices as directed to maintain accuracy.

- **Review data with a healthcare provider**: Share device data with healthcare providers during appointments to receive informed feedback on adjustments.
- **Use trend data, not just instant readings:** Focus on trends and patterns over time rather than individual readings. This broader view provides a better understanding of overall glucose control.
- **Customize alerts thoughtfully:** Set CGM alerts to manageable levels that provide adequate warning without causing alarm fatigue.
- **Backup and save data:** Periodically backup device data to prevent loss and track long-term progress.
- **Keep supplies and batteries available:** Maintain spare supplies for all devices, including sensors, infusion sets, batteries, or chargers, to avoid disruptions.
- **Stay updated with device maintenance:** Regular maintenance, software updates, and cleaning ensure devices function optimally and reduce the risk of malfunctions.
- **Educate yourself on device features:** Take time to understand all features of each feature.

Conclusion

The integration of technology into diabetes management has ushered in a new era of healthcare, significantly improving the lives of those living with diabetes. The advancements in continuous glucose monitors (CGMs), insulin pumps, and smartphone apps have made it possible to monitor and manage blood sugar levels with unprecedented precision and ease. These technologies not only provide real-time data but also facilitate better communication and collaboration between patients and healthcare providers, leading to more personalized and effective treatment plans.

The rise of telemedicine has further expanded access to diabetes care, allowing individuals to consult with their healthcare providers remotely. This development is particularly beneficial for those living in rural areas or with mobility issues, ensuring that they receive timely and consistent care. Wearable devices, such as smartwatches, enhance this ecosystem by integrating various health metrics, from physical activity to heart rate, into a cohesive diabetes management plan.

Emerging technologies like artificial intelligence (AI) and virtual reality (VR) hold great promise for the future of diabetes care. AI can analyze vast amounts of data to predict blood sugar trends and provide tailored recommendations, while VR can create immersive educational experiences that help individuals better understand and manage their condition. These innovations are set to make diabetes management more intuitive and effective.

Despite the numerous benefits, it is important to acknowledge and address the challenges associated with diabetes technology. The cost of devices, technical issues, and data security concerns are significant factors that need careful consideration. By choosing the right technology, maintaining devices properly, and protecting personal health data, individuals can maximize the benefits of these tools while minimizing potential drawbacks.

Ultimately, the future of diabetes management is bright, with technology playing a central role in improving health outcomes and quality of life. By staying informed about the latest advancements and integrating them into daily routines, individuals with diabetes can achieve better control over their condition, reduce the risk of complications, and lead healthier, more fulfilling lives. The ongoing collaboration between patients, healthcare providers, and technology developers will continue to drive innovation, making diabetes management more accessible and effective for all.

Bibliography

1. American Diabetes Association. (n.d.). *Diabetes technology guide.* https://diabetes.org/living-with-diabetes/treatment-care/diabetes-technology-guide
2. Diabetes Canada. (n.d.). *Technology & devices.* https://diabetes.ca/about-diabetes/diabetes-management/tech-devices
3. American Diabetes Association. (n.d.). *Devices and technology.* https://diabetes.org/about-diabetes/devices-technology
4. American Diabetes Association Professional Practice Committee. (2024, January). Diabetes technology: Standards of care in diabetes—2024. *Diabetes Care, 47*(Supplement 1), S126–S144. https://doi.org/10.2337/dc24-S007
5. Endocrine Society. (2022, January 24). *Diabetes technology.* https://www.endocrine.org/patient-engagement/endocrine-library/diabetes-technology
6. Edelman, S., Cheatham, W. W., Norton, A., & Close, K. L. (2024, Spring). Patient perspectives on the benefits and challenges of diabetes and digital technology. *Clinical Diabetes, 42*(2), 243–256. https://doi.org/10.2337/cd23-0003
7. Zahid, M., Dowlatshahi, S., Kansara, A. H., & Sadhu, A. R. (2023, August). The evolution of diabetes technology – options toward personalized care. *Endocrine Practice, 29*(8), 653–662. https://doi.org/10.1016/j.eprac.2023.04.007
8. Ebekozien, O., Fantasia, K., Farrokhi, F., Sabharwal, A., & Kerr, D. (2024, January 31). Technology and health inequities in diabetes care: How do we widen access to underserved populations and utilize technology to improve outcomes for all? *Diabetes, Obesity & Metabolism, 26 Suppl 1*(Suppl 1), 3–13. https://doi.org/10.1111/dom.15470

CHAPTER 13

Diabetes and Financial Management

Introduction

Managing diabetes effectively goes beyond just medical care; it encompasses a broad range of financial considerations that can significantly impact an individual's well-being. Chapter 13 , "Diabetes and Financial Management" delves into the essential strategies and resources available to help people with diabetes manage the financial aspects of their condition. This chapter addresses various questions and concerns, providing practical advice on how to handle the costs associated with diabetes care.

The cost of medications and supplies can be daunting, but there are ways to alleviate these expenses, such as opting for generic options, using discount programs, and exploring patient assistance programs. Effective budgeting and understanding insurance coverage are crucial for managing medical expenses. This chapter offers insights on how to navigate insurance claims, reduce out-of-pocket costs, and plan for unexpected medical expenses. It also highlights the importance of building an emergency fund and utilizing Health Savings Accounts (HSAs).

For those facing job loss or retirement, the chapter provides guidance on securing financial assistance, managing healthcare costs, and planning for the future. Additionally, it discusses strategies for reducing costs related to healthy eating, exercise, vision and dental care, and mental health services. Practical tips on finding affordable healthcare providers, managing medical debt, and handling insurance denials are also covered.

By implementing the strategies outlined in this chapter, individuals with diabetes can achieve a more secure financial footing, ensuring that they can focus on their health and well-being without being overwhelmed by financial stress. Whether you are newly diagnosed or have been managing diabetes for years, the information provided here will equip you with the knowledge and tools needed to navigate the financial challenges of living with diabetes.

1. How can people with diabetes manage the cost of medications?

Look for generic options, use prescription discount programs, and explore patient assistance programs.

2. How can people with diabetes save money on diabetes supplies?

Buy in bulk, use mail-order pharmacies, and seek out discounts and coupons.

3. What are some financial assistance programs for people with diabetes?

Programs include Medicaid, Medicare, state assistance programs, and nonprofit organizations.

4. How can people with diabetes budget for their medical expenses?

Track expenses, prioritize needs, and set aside funds for healthcare costs.

5. How can health insurance help cover diabetes-related costs?

Health insurance can cover medications, supplies, doctor visits, and hospital stays.

6. How can people with diabetes navigate insurance claims?

Keep detailed records, understand insurance policies, and work with healthcare providers on documentation.

7. What are some tips for reducing out-of-pocket healthcare costs?

Use in-network providers, negotiate medical bills, and explore preventive care options.

8. How can people with diabetes manage their finances during job loss?

Apply for unemployment benefits, explore COBRA insurance, and seek temporary financial assistance.

9. How can people with diabetes plan for retirement?

Save regularly, consider healthcare costs, and explore long-term care insurance options.

10. How can people with diabetes handle medical debt?

Negotiate payment plans, seek financial counseling, and explore debt relief programs.

11. How can people with diabetes reduce the cost of healthy eating?

Plan meals, buy seasonal produce, and use coupons and sales.

12. How can people with diabetes save money on exercise and fitness?

Use free resources like parks and online workouts and seek out community fitness programs.

13. How can people with diabetes find affordable healthcare providers?

Use community health centers, explore telemedicine options, and seek recommendations.

14. How can people with diabetes prepare for unexpected medical expenses?

Build an emergency fund, use health savings accounts (HSAs), and have a financial plan in place.

15. How can people with diabetes handle financial stress?

Practice stress- reduction techniques, seek financial counseling, and focus on budgeting.

16. How can people with diabetes save money on vision care?

Use vision insurance, seek out discount programs, and look for affordable eye care providers.

17. How can people with diabetes save money on dental care?

Use dental insurance, seek out dental schools, and explore discount dental plans.

18. How can people with diabetes find affordable mental health services?

Use mental health insurance benefits, seek out community resources, and explore sliding scale fees.

19. How can people with diabetes manage healthcare costs while on a fixed income?

Budget carefully, seek out financial assistance programs, and prioritize healthcare needs.

20. How can people with diabetes use tax benefits for medical expenses?

Deduct eligible medical expenses, use health savings accounts (HSAs), and explore tax credits.

21. How can people with diabetes plan for long-term care?

Consider long-term care insurance, save for future expenses, and explore state and federal assistance programs.

22. How can people with diabetes handle insurance denials?

Appeal the decision, provide additional documentation, and seek assistance from healthcare providers.

23. How can people with diabetes find affordable medications?

Use prescription discount cards, seek out pharmaceutical assistance programs, and consider mail-order pharmacies.

24. How can people with diabetes reduce the cost of blood sugar monitoring?

Use cost-effective testing supplies, seek out discounts, and monitor usage.

25. How can people with diabetes find affordable medical devices?

Explore insurance coverage, seek out discount programs, and buy refurbished devices.

26. How can people with diabetes save money on transportation for medical appointments?

Use public transportation, seek out ride-sharing options, and explore community transportation programs.

27. How can people with diabetes manage the cost of home modifications?

Seek out grants and loans, use insurance benefits, and explore nonprofit programs.

28. How can people with diabetes find affordable health insurance?

Compare plans, seek out subsidies, and explore state and federal insurance marketplaces.

29. How can people with diabetes reduce the cost of diabetes education programs?

Use insurance benefits, seek out community programs, and explore online resources.

30. How can people with diabetes save money on over-the-counter medications?

Buy generic options, use store brands, and seek out discounts and coupons.

31. How can people with diabetes find financial counseling?

Seek out nonprofit organizations, use community resources, and explore online financial counseling services.

32. How can people with diabetes manage healthcare costs during a crisis?

Seek out emergency financial assistance, use savings, and prioritize healthcare needs.

33. How can people with diabetes find affordable housing?

Explore housing assistance programs, seek out affordable housing options, and use community resources.

34. How can people with diabetes manage healthcare costs during travel?

Use travel insurance, plan for medical expenses, and carry necessary supplies.

35. How can people with diabetes handle rising healthcare costs?

Budget carefully, seek out financial assistance programs, and advocate for policy changes.

36. How can people with diabetes save money on medical tests?

Use in-network providers, seek out discount programs, and ask about cost-effective alternatives.

37. How can people with diabetes reduce the cost of wound care?

Use insurance benefits, seek out discount programs, and follow preventive care guidelines.

38. How can people with diabetes manage the cost of insulin?

Use prescription discount programs, seek out patient assistance programs, and consider biosimilar options.

39. How can people with diabetes find affordable medical supplies?

Use bulk purchasing, seek out discounts, and explore nonprofit assistance programs.

40. How can people with diabetes save money on foot care?

Use insurance benefits, seek out community resources, and follow preventive care guidelines.

41. How can people with diabetes manage the cost of hearing care?

Use insurance benefits, seek out discount programs, and explore community resources.

42. How can people with diabetes handle unexpected medical bills?

Negotiate payment plans, seek financial counseling, and explore emergency financial assistance.

43. How can people with diabetes save money on alternative therapies?

Use insurance benefits, seek out community resources, and explore discount programs.

44. How can people with diabetes reduce the cost of hospital stays?

Use in-network hospitals, seek out financial assistance programs, and negotiate bills.

45. How can people with diabetes find affordable health and wellness programs?

Use community resources, seek out discount programs, and explore online options.

46. How can people with diabetes manage healthcare costs while caregiving?

Budget carefully, seek out financial assistance programs, and prioritize healthcare needs.

47. How can people with diabetes save money on prescription eyeglasses?

Use vision insurance benefits, seek out discount programs, and consider online retailers.

48. How can people with diabetes manage the cost of emergency care?

Use insurance benefits, seek out financial assistance programs, and negotiate bills.

49. How can people with diabetes find affordable mental health medications?

Use prescription discount programs, seek out patient assistance programs, and explore generic options.

Use insurance benefits, seek out community resources, and prioritize preventive care appointments.

Conclusion

Navigating the financial landscape of diabetes care can be daunting, but with informed strategies and proactive approaches, individuals can effectively manage their healthcare expenses while maintaining their health and well-being. From medication costs to preventive care, each aspect of managing diabetes-related expenses requires careful planning and resourcefulness.

A cornerstone of managing costs is understanding available resources and utilizing them effectively. Generic medications, prescription discount programs, and patient assistance programs offer significant savings on medications, ensuring access to essential treatments without financial strain. Similarly, purchasing diabetes supplies in bulk, exploring mail-order options, and utilizing discounts and coupons can substantially reduce expenses associated with daily management.

Financial assistance programs, including Medicaid, Medicare, state assistance programs, and support from nonprofit organizations, provide vital resources for those facing financial challenges. These programs not only alleviate immediate financial burdens but also ensure continued access to necessary healthcare services and medications.

Budgeting and financial planning play pivotal roles in managing diabetes-related expenses. Tracking medical expenses, prioritizing healthcare needs, and setting aside funds for healthcare costs contribute to financial stability and peace of mind. Health insurance coverage, including provisions for medications, supplies, and preventive care, offers essential protection against unexpected healthcare expenses.

Effective navigation of insurance claims, including maintaining detailed records and understanding policy coverage, helps individuals maximize their insurance benefits and minimize out-of-pocket costs. When facing insurance denials or challenges, knowing how to appeal decisions and advocate for coverage can make a significant difference in accessing necessary treatments and services.

Moreover, proactive health management, including preventive care and regular checkups, not only reduces long-term healthcare costs but also promotes overall well-being. By prioritizing preventive measures and adopting a proactive approach to healthcare, individuals can mitigate the risk of complications and associated costs over time.

In conclusion, managing healthcare costs for diabetes requires a multifaceted approach that combines financial savvy with proactive healthcare management. By leveraging available resources, staying informed about financial assistance options, and adopting prudent financial practices, individuals can navigate the financial challenges of diabetes care while focusing on achieving optimal health outcomes and quality of life. Empowered with knowledge and proactive strategies, individuals can confidently manage their diabetes-related expenses and maintain a balanced approach to health and finances.

Bibliography

1. Herges, J. R., Neumiller, J. J., & McCoy, R. G. (2021, October). Easing the Financial Burden of Diabetes Management: A Guide for Patients and Primary Care Clinicians. *Clinical diabetes: a publication of the American Diabetes Association, 39*(4), 427–436. https://doi.org/10.2337/cd21-0004
2. Parker, E. D., Lin, J., Mahoney, T., Ume, N., Yang, G., Gabbay, R. A., ElSayed, N. A., & Bannuru, R. R. (2024, January). Economic costs of diabetes in the U.S. in 2022. *Diabetes Care, 47*(1), 26–43. https://doi.org/10.2337/dci23-0085
3. Diabetes Research Connection. (2024, January 16). *Managing diabetes on a budget.* https://diabetesresearchconnection.org/managing-diabetes-on-a-budget/
4. National Institute of Diabetes and Digestive and Kidney Diseases. (2019, June). *Financial help for diabetes care.* https://www.niddk.nih.gov/health-information/diabetes/financial-help-diabetes-care
5. American Diabetes Association (n.d.). *4 easy ways to help manage diabetes care costs.* https://diabetes.org/diabetes-financial-impact/manage-diabetes-care-costs
6. American Society for the Positive Care of Children. (2024, May 15). *Understanding the financial landscape of diabetes: A guide for families.* https://americanspcc.org/understanding-the-financial-landscape-of-diabetes-a-guide-for-families/
7. Caraballo, C., Valero-Elizondo, J., Khera, R., Mahajan, S., Grandhi, G. R., Virani, S. S., Mszar, R., Krumholz, H. M., & Nasir, K. (2020, February 6). Burden and consequences of financial hardship from medical bills among nonelderly adults with diabetes mellitus in the United States. *Circulation: Cardiovascular Quality and Outcomes, 13*(2), e006139. https://doi.org/10.1161/CIRCOUTCOMES.119.006139

CHAPTER 14

A Holistic Approach to Diabetes Management

Introduction

Living well with diabetes is not just about managing blood sugar levels; it encompasses a holistic approach that integrates physical, emotional, and social well-being into daily life. This comprehensive guide explores essential strategies and principles to empower individuals with diabetes to thrive and maintain optimal health.

Central to living well with diabetes is effective blood sugar management. By understanding the impact of carbohydrates, proteins, fats, and fiber on blood glucose levels, individuals can make informed dietary choices that support stable blood sugar control. Coupled with regular physical activity tailored to individual abilities and preferences, maintaining a healthy lifestyle becomes a cornerstone of diabetes management.

Beyond the physical aspects, emotional well-being plays a crucial role. Coping with the daily challenges of diabetes requires resilience, mindfulness, and strategies to manage stress and emotional fluctuations effectively. Building a strong support network, involving loved ones in diabetes care, and seeking professional guidance contribute to emotional resilience and overall mental health.

Engaging with a diabetes care team, which may include healthcare providers, nutritionists, and mental health professionals, provides comprehensive support and guidance tailored to individual needs. Regular checkups, screenings, and staying informed about the latest developments in diabetes management ensure proactive health management and early intervention when necessary.

Moreover, technology plays a pivotal role in modern diabetes care, offering tools such as glucose monitors, insulin pumps, and mobile apps that aid in tracking blood sugar levels, managing medications, and accessing educational resources.

Throughout this exploration, we delve into practical strategies for setting realistic health goals, maintaining motivation, and navigating life transitions while managing diabetes. By embracing a proactive and multifaceted approach to diabetes care, individuals can not only manage their condition effectively but also lead fulfilling and active lives.

This guide aims to empower individuals with diabetes with knowledge, tools, and inspiration to achieve optimal health and well-being, fostering a positive and proactive mindset toward living well with diabetes.

1. What are the key components of living well with diabetes?

Key components include effective blood sugar management, healthy lifestyle choices, regular medical care, and emotional well-being.

2. How can people with diabetes set realistic health goals?

Set specific, measurable, achievable, relevant, and time-bound (SMART) goals.

3. Why is it important to have a diabetes care team?

A care team provides comprehensive support, including medical, nutritional, and emotional care.

4. How can people with diabetes stay informed about their condition?

Follow reputable sources, attend diabetes education programs, and engage with healthcare providers.

5. What role does technology play in diabetes management?

Technology, such as glucose monitors and apps, helps track blood sugar levels, manage medications, and provide educational resources.

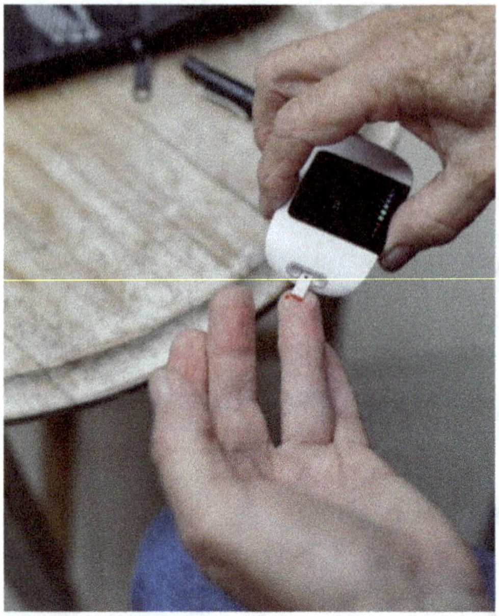

▲ Figure 14.1
Glucose monitor.

6. How can people with diabetes balance self-care with daily responsibilities?

Integrate self-care into routines, prioritize tasks, and seek support when needed.

7. What are some strategies for preventing diabetes complications?

Maintain good blood sugar control, follow medical advice, and adopt a healthy lifestyle.

8. How can people with diabetes stay motivated over the long term?

Set new goals, celebrate achievements, and seek ongoing support and education.

9. What is the importance of regular checkups and screenings?

Regular checkups help detect and manage complications early, ensuring better long-term health outcomes.

10. How can people with diabetes advocate for themselves in healthcare settings?

Communicate clearly, ask questions, and participate actively in treatment decisions.

11. What are some ways to manage diabetes during life transitions?

Plan ahead, adjust management strategies, and seek support from healthcare providers and loved ones.

12. How can people with diabetes cultivate a positive mindset?

Practice gratitude, focus on strengths, and engage in activities that bring joy and fulfillment.

13. How can people with diabetes involve their family in their care?

Educate family members, communicate needs and preferences, and involve them in management tasks and support.

14. What are some tips for maintaining a healthy work–life balance with diabetes?

Set boundaries, prioritize self-care, and use time management techniques.

15. How can people with diabetes find support in their communities?

Join local diabetes support groups, participate in community events, and connect with online communities.

16. What are some benefits of volunteering or helping others for people with diabetes?

Volunteering can provide a sense of purpose, reduce stress, and improve overall well-being.

17. How can people with diabetes handle setbacks in their management of the disease?

View setbacks as learning opportunities, seek support, and adjust strategies as needed.

18. What are some resources for people with diabetes seeking financial assistance?

Look into patient assistance programs, nonprofits, and government resources for financial support.

19. How can people with diabetes stay active in their communities?

Participate in local events, join clubs or organizations, and engage in social activities.

20. What are some ways to celebrate achievements in diabetes management?

Reward yourself with nonfood-related treats, share successes with others, and reflect on progress.

21. How can people with diabetes use goal-setting to improve their health?

Set clear, attainable goals, track progress, and adjust as needed to stay motivated.

22. How can people with diabetes practice self-compassion?

Treat yourself with kindness, recognize your efforts, and avoid self-criticism.

23. What is the importance of finding balance in diabetes management?

Balance helps prevent burnout, maintain motivation, and improve overall well-being.

24. How can people with diabetes integrate physical activity into their daily lives?

Find activities you enjoy, set a regular schedule, and involve friends or family for support.

25. How can people with diabetes handle stress related to their condition?

Practice stress-management techniques, seek support, and maintain a healthy lifestyle.

26. How can people with diabetes improve their mental health?

Seek therapy or counseling, practice mindfulness, and engage in activities that promote relaxation and joy.

27. What are some strategies for managing diabetes in social settings?

Plan ahead, educate others about diabetes, and stay flexible with management strategies.

28. How can people with diabetes build a strong support network?

Connect with healthcare providers, join support groups, and communicate with friends and family.

29. How can people with diabetes maintain hope and optimism?

Focus on progress, set achievable goals, and stay connected with supportive communities.

30. What is the importance of self-education in diabetes management?

Staying informed helps people make better decisions, stay motivated, and effectively manage their condition.

31. How can people with diabetes handle emotions related to their condition?

Acknowledge feelings, seek support, and use coping strategies to manage emotions.

32. How can people with diabetes find joy and fulfillment in life?

Pursue hobbies, build strong relationships, and focus on positive experiences.

33. How can people with diabetes manage their condition during travel?

Plan ahead, pack necessary supplies, and stay flexible with management strategies.

34. What are some tips for preventing diabetes burnout?

Take breaks, seek support, adjust management strategies, and prioritize self-care.

35. How can people with diabetes practice mindfulness?

Engage in meditation, mindful eating, and mindful movement practices like yoga.

36. What is the role of pets in supporting people with diabetes?

Pets provide companionship, reduce stress, and encourage physical activity.

37. How can people with diabetes balance independence and seeking help?

Recognize when support is needed, communicate needs clearly, and maintain a sense of autonomy.

38. What are some benefits of nature for people with diabetes?

Spending time in nature can reduce stress, improve mood, and promote physical activity.

39. How can people with diabetes handle negative emotions related to their condition?

Acknowledge emotions, seek support, and focus on positive aspects of life.

40. How can people with diabetes develop resilience?

Build a support network, focus on problem-solving skills, and maintain a positive outlook.

41. How can people with diabetes use art or creative activities for emotional well-being?

Creative activities can provide an outlet for expression, reduce stress, and improve mood.

42. What are some tips for managing diabetes in different life stages?

Adjust management strategies, seek support from healthcare providers, and stay informed about changes.

43. How can people with diabetes stay motivated over the long term?

Set new goals, celebrate achievements, and seek ongoing support and education.

44. How can people with diabetes balance self-care with daily responsibilities?

Integrate self-care into routines, prioritize tasks, and seek support when needed.

45. What are some benefits of volunteering for people with diabetes?

Volunteering can provide a sense of purpose, reduce stress, and improve overall well-being.

46. How can people with diabetes find support in their communities?

Join local diabetes support groups, participate in community events, and connect with online communities.

47. What are some strategies for maintaining a healthy work-life balance with diabetes?

Set boundaries, prioritize self-care, and use time management techniques.

48. How can people with diabetes involve their family in their care?

Educate family members, communicate needs and preferences, and involve them in management tasks and support.

49. How can people with diabetes develop a positive mindset?

Practice gratitude, focus on strengths, and engage in activities that bring joy and fulfillment.

50. What are some ways to celebrate achievements in diabetes management?

Reward yourself with nonfood-related treats, share successes with others, and reflect on progress.

Conclusion

In conclusion, living well with diabetes requires a holistic approach that encompasses effective blood sugar management, healthy lifestyle choices, regular medical care, and nurturing emotional well-being. Throughout this exploration of essential aspects of diabetes management, several foundational principles have emerged, each crucial for maintaining optimal health and quality of life.

First and foremost, effective blood sugar management stands as the cornerstone of diabetes care. This involves monitoring blood glucose levels consistently, adhering to prescribed medications or insulin regimens, and making informed dietary choices to maintain stable blood sugar levels throughout the day. Integrating physical activity into daily routines further enhances blood sugar control, improves cardiovascular health, and boosts overall well-being.

Beyond the physiological aspects, adopting a healthy lifestyle plays a pivotal role in diabetes management. This includes maintaining a balanced diet rich in whole grains, lean proteins, healthy fats, and fiber-rich foods like fruits and vegetables. Portion control and mindful eating practices help individuals with diabetes manage their weight and optimize blood sugar levels.

Regular medical care is essential for ongoing support and management of diabetes. Building a diabetes care team that includes healthcare professionals such as endocrinologists, dietitians, and diabetes educators ensures comprehensive care tailored to individual needs. Engaging in regular checkups, screenings, and diabetes education programs empowers individuals to stay informed about advancements in treatment and management strategies.

Emotional well-being is equally vital in diabetes management. Cultivating a positive mindset, practicing self-compassion, and managing stress effectively contribute to overall mental health. Seeking support from family, friends, and diabetes support groups can provide invaluable encouragement and motivation on the journey.

Looking ahead, setting realistic health goals using SMART criteria, staying informed through reputable sources, and leveraging technology for tracking progress and accessing resources further supports long-term success in managing diabetes.

In essence, by embracing these principles and strategies, individuals with diabetes can lead fulfilling lives, minimize the risk of complications, and navigate the challenges of diabetes with resilience and optimism.

This conclusion summarizes the importance of each component of diabetes management, highlights strategies for success, and emphasizes the importance of a comprehensive approach to living well with diabetes.

Bibliography

1. American Heart Association. (2024, April 4). *Living healthy with diabetes.* https://www.heart.org/en/health-topics/diabetes/prevention--treatment-of-diabetes/living-healthy-with-diabetes
2. Centers for Disease Control and Prevention. (2024, May 15). *Living with diabetes.* https://www.cdc.gov/diabetes/living-with/index.html
3. American Diabetes Association Professional Practice Committee, ElSayed, N. A., Aleppo, G., Bannuru, R. R., Beverly, E. A., Bruemmer, D., Collins, B. S., Darville, A., Ekhlaspour, L., Hassanein, M., Hilliard, M. E., Johnson, E. L., Khunti, K., Lingvay, I., Matfin, G., McCoy, R. G., Perry, M. L., Pilla, S. J., Polsky, S., ... Gabbay, R. A. (2024, January). 5. Facilitating positive health behaviors and well-being to improve health outcomes: Standards of care in diabetes—2024. *Diabetes Care, 47*(Supplement_1), S77–S110. https://doi.org/10.2337/dc24-S005
4. National Institute of Diabetes and Digestive and Kidney Diseases. (2023, October). *Healthy living with diabetes.* https://www.niddk.nih.gov/health-information/diabetes/overview/healthy-living-with-diabetes
5. Mayo Clinic. (2022, April 19). *How to live well with diabetes.* https://www.mayoclinichealthsystem.org/hometown-health/speaking-of-health/how-to-live-well-with-diabetes
6. DiLonardo, M. J. (2025, April 14). *6 lifestyle changes to control your diabetes.* WebMD. https://www.webmd.com/diabetes/diabetes-lifestyle-tips
7. Juanamasta, I. G., Aungsuroch, Y., Gunawan, J., Suniyadewi, N. W., & Nopita Wati, N. M. (2021). Holistic care management of diabetes mellitus: An integrative review. *International Journal of Preventive Medicine, 12*, 69. https://pubmed.ncbi.nlm.nih.gov/34447511/

Research and Future Treatments for Diabetes

CHAPTER 15

Introduction

Diabetes mellitus, a chronic metabolic disorder characterized by persistent hyperglycemia, poses significant challenges to global health, affecting millions of individuals worldwide. The disease is primarily classified into two main types: Type 1 diabetes, which results from autoimmune destruction of insulin-producing beta cells, and Type 2 diabetes, which is often associated with insulin resistance and influenced by lifestyle factors and genetics. The increasing prevalence of diabetes necessitates a comprehensive understanding of its pathophysiology, management strategies, and the development of innovative treatments aimed at improving patient outcomes and quality of life.

Research has highlighted the importance of self-management and patient education in diabetes care. Effective self-care practices, such as adherence to dietary modifications and regular physical activity, are crucial for managing blood glucose levels and preventing complications. Studies indicate that culturally tailored interventions can significantly enhance understanding and implementation of diabetes self-management among diverse populations, including Native Hawaiians and Pacific Islanders. Furthermore, the role of healthcare professionals, particularly nurses, in facilitating patient education and support is paramount, as their understanding of diabetes directly impacts patient care and outcomes.

Emerging treatments for diabetes are increasingly focusing on the underlying mechanisms of the disease. For instance, recent advancements in understanding the role of oxidative stress in diabetic neuropathy have opened avenues for potential therapeutic interventions targeting this pathway. Additionally, the exploration of mitochondrial dysfunction as a therapeutic target represents a promising frontier in diabetes treatment. Furthermore, the bidirectional relationship between diabetes and periodontal disease underscores the need for integrated care approaches that address both metabolic and oral health. As research continues to evolve, the future of diabetes treatment may involve a more personalized approach, utilizing biomarkers and genetic profiling to tailor interventions to individual patient needs. This shift toward precision medicine aims to enhance the efficacy of treatments and improve overall management of diabetes, ultimately leading to better health outcomes for individuals living with this chronic condition.

In conclusion, understanding diabetes requires a multifaceted approach that encompasses patient education, innovative research, and the development of targeted therapies. By fostering a deeper comprehension of the disease and its management, we can pave the way for improved treatment strategies and enhanced quality of life for those affected by diabetes.

1. What is artificial pancreas technology?

Artificial pancreas technology, also known as a closed-loop insulin delivery system, is a breakthrough diabetes management tool designed to automate blood glucose control for people with diabetes, particularly Type 1 diabetes. This technology mimics the glucose-regulating function of a healthy pancreas by combining three main components: a continuous glucose monitor (CGM), an insulin pump, and an algorithm that interprets data and adjusts insulin delivery.

2. What are the latest advancements in diabetes treatment research?

Diabetes research is progressing rapidly, with scientists and clinicians exploring new ways to improve management, delay onset, and ultimately find a cure. Some recent advancements include:

- **Beta-cell regeneration and replacement:** Researchers are studying methods to stimulate the body to regenerate insulin-producing beta cells, which are destroyed in Type 1 diabetes or become dysfunctional in Type 2. This could allow the body to restore its natural insulin production.
- **Islet cell transplants:** Islet cell transplantation involves transplanting insulin-producing cells from a donor into a person with diabetes. Studies show that some recipients achieve insulin independence, though immunosuppression requirements remain a challenge.

- **Advanced insulin formulations:** New forms of ultra-rapid insulin have been developed that begin working within minutes and mimic the body's natural insulin response more closely. Researchers are also working on "smart insulin " that activates in response to blood glucose levels.
- **Closed-loop systems:** Often referred to as the artificial pancreas, closed-loop systems combine continuous glucose monitors (CGMs) and insulin pumps to automate insulin delivery based on real-time glucose data, reducing the need for manual intervention.
- **Diabetes vaccine research:** For Type 1 diabetes, studies are exploring vaccines to halt the immune attack on beta cells. One potential vaccine targets the immune cells responsible for destroying insulin-producing cells in genetically predisposed individuals.

These advancements are helping improve daily diabetes management and offer hope for treatments that could reduce the need for insulin injections or other lifelong therapies.

3. How could gene therapy potentially revolutionize Type 1 diabetes treatment?

Gene therapy has the potential to be a game-changer in treating Type 1 diabetes, which is fundamentally an autoimmune disease where the body's immune system attacks insulin-producing beta cells. Current gene therapy research focuses on several strategies:

- **Modifying beta cells to resist autoimmune attack:** Scientists are exploring ways to alter the genes of beta cells to make them invisible to the immune system or resistant to autoimmune destruction. By preserving beta cell function, gene therapy could restore the body's ability to produce insulin naturally.
- **Beta cell regeneration via gene editing:** Using tools like **Clustered Regularly Interspaced Short Palindromic Repeats** (CRISPR), researchers can potentially edit genes that control beta cell regeneration and function. CRISPR is a powerful and precise gene-editing technology. Comparatively it is a "find and replace" tool or a pair of "molecular scissors" for DNA.

This could lead to therapies that regenerate beta cells lost to the autoimmune attack.

- **Immunomodulation:** Another gene therapy approach involves modifying immune cells to stop them from attacking beta cells. By "retraining" the immune system, it may be possible to prevent Type 1 diabetes progression.
- **In vivo production of insulin:** Gene therapy can potentially reprogram other types of cells (like liver cells) to produce insulin, effectively replacing the lost beta cells. This method could enable the body to produce insulin again without relying on pancreatic beta cells.

Although still in experimental stages, gene therapy holds promise as a one-time treatment that could correct the underlying genetic and autoimmune issues, offering a possible cure for Type 1 diabetes.

4. What are the promising developments in immune therapy for Type 1 diabetes?

Immune therapy is a promising area for Type 1 diabetes, targeting the root cause of the disease: the immune system's attack on insulin-producing beta cells. Key developments include:

- **Immune tolerance therapy:** Researchers are developing therapies that retrain the immune system to tolerate beta cells instead of attacking them. This approach aims to stop or slow the progression of Type 1 diabetes by halting the autoimmune response.
- **Anti-CD3 antibodies:** Medications like teplizumab, an anti-CD3 monoclonal antibody, have shown promise in delaying the onset of Type 1 diabetes in individuals at high risk. Teplizumab works by modifying the immune response, preserving beta cell function and delaying the disease's progression.
- **T regulatory cells (Tregs):** These cells play a key role in controlling the immune response. By enhancing the function or number of Tregs, researchers hope to prevent the immune system from attacking beta cells. Treg therapy is showing potential in early-stage clinical trials.

- **Nanoparticle-based immunotherapy:** Some researchers are using nanoparticles to deliver immune-modulating agents directly to pancreatic islets. This localized therapy aims to suppress the immune response against beta cells without causing systemic immunosuppression.
- **Peptide and protein therapies:** Specific proteins or peptides can help induce immune tolerance to insulin-producing cells. By exposing the immune system to beta cell proteins in a controlled way, it may be possible to retrain immune cells to recognize beta cells as part of the body rather than foreign invaders.

These immune therapies have the potential to change Type 1 diabetes from an actively managed condition to one where progression is halted or even reversed.

5. How are new medications impacting the prevention of Type 2 diabetes?

Preventing Type 2 diabetes is increasingly feasible due to new medications that target risk factors like insulin resistance, obesity, and elevated blood sugar levels. Some of the key developments include:

- **GLP-1 receptor agonists:** These medications, such as liraglutide and semaglutide, help reduce blood glucose levels and promote weight loss. They work by stimulating insulin secretion and reducing appetite. For people with prediabetes or high diabetes risk, GLP-1 receptor agonists are proving effective in lowering the risk of developing Type 2 diabetes.
- **SGLT2 inhibitors:** These drugs, including canagliflozin and dapagliflozin, lower blood glucose by promoting glucose excretion in the urine. Studies show that SGLT2 inhibitors also have cardiovascular and renal benefits, making them ideal for people with metabolic syndrome or high-risk profiles.
- **Metformin for prediabetes:** Metformin remains a cornerstone in preventing diabetes for people with prediabetes, especially in individuals with high BMI or a family history of diabetes. It improves insulin sensitivity and reduces hepatic glucose production.

- **Newer insulin sensitizers:** Researchers are investigating new insulin-sensitizing drugs, like thiazolidinediones (TZDs) and other PPAR agonists, to address insulin resistance more effectively and potentially reduce diabetes onset.
- **Combination therapy for high-risk individuals:** Combining metformin, GLP-1 receptor agonists, and SGLT2 inhibitors is being studied to prevent Type 2 diabetes progression in people at high risk.

These medications are helping individuals manage their prediabetes more effectively and reduce the overall risk of developing Type 2 diabetes by addressing its root causes.

6. What advancements are being made in artificial pancreas technology?

The artificial pancreas, also known as a closed-loop system, is a breakthrough in diabetes technology that automates blood glucose management by combining continuous glucose monitoring (CGM) and insulin delivery. Recent advancements in artificial pancreas technology include:

- **Hybrid closed-loop systems:** Most commercial artificial pancreas systems are hybrid, meaning they automate basal insulin delivery but require users to manually input bolus insulin doses for meals. These systems, such as the Tandem Control-IQ and Medtronic MiniMed 670G, have shown great success in improving time-in-range and reducing A1C levels.
- **Fully automated closed-loop systems:** Fully automated systems are under development and would adjust both basal and bolus insulin doses without user input. These systems use advanced algorithms to predict and respond to blood glucose changes, enabling hands-free management.
- **Dual-hormone artificial pancreas:** Some systems under research include both insulin and glucagon delivery. This dual-hormone approach better mimics the natural hormonal balance and may reduce the risk of hypoglycemia during intense physical activities or overnight.

- **Interoperability and customization:** New artificial pancreas systems are becoming more interoperable, allowing users to pair their CGM and insulin pump devices from different manufacturers. This customization gives users more flexibility in selecting devices that fit their lifestyle.
- **Remote monitoring and control:** Some systems allow for remote monitoring and control through smartphone apps, providing caregivers and healthcare providers with the ability to adjust settings and monitor trends from afar.

Advancements in artificial pancreas technology are making diabetes management simpler, safer, and more effective, with fewer daily interruptions.

7. How could wearable technology change the future of diabetes management?

Wearable technology is shaping the future of diabetes management by providing real-time health data, enabling more proactive care, and enhancing quality of life. The important ways wearable tech could transform diabetes management include:

- **Continuous health monitoring:** Wearables, such as smartwatches and fitness bands, can continuously track metrics like heart rate, physical activity, and sleep patterns. Integrated with CGM data, these metrics offer a comprehensive view of factors influencing blood glucose levels.
- **Smart insulin pens and insulin delivery patches:** These devices log and track insulin doses, providing reminders and historical data for better dosing accuracy. Insulin patches deliver continuous, low doses of insulin without injections and may integrate with CGMs in the future.
- **Real-time alerts and notifications:** Wearables provide real-time alerts for high or low blood glucose, making it easier for users to take immediate action. This is particularly helpful for individuals who may not feel symptoms of hypo- or hyperglycemia.
- **Data sharing and telemedicine integration:** Wearables can transmit glucose data to healthcare

providers in real-time, enabling telemedicine support and remote adjustments to treatment plans. This feature is invaluable for those in remote areas or who need frequent monitoring.

- **Personalized insights and AI:** Artificial intelligence (AI) algorithms analyze data collected from wearables to provide.

Conclusion

In conclusion, the multifaceted nature of diabetes mellitus necessitates a comprehensive approach to research and treatment that encompasses both the biological underpinnings of the disease and the psychosocial factors influencing patient management. As the global prevalence of diabetes continues to rise, understanding the complexities of this condition becomes increasingly critical. Current research highlights the importance of effective self-management strategies, patient education, and culturally sensitive interventions that empower individuals to take control of their health.

Furthermore, advancements in our understanding of the pathophysiology of diabetes, including oxidative stress and mitochondrial dysfunction, pave the way for innovative therapeutic approaches that may enhance treatment efficacy and patient outcomes. The future of diabetes care is poised to embrace precision medicine, tailoring interventions based on individual patient characteristics, genetic profiles, and specific disease mechanisms. This shift toward personalized treatment strategies holds the promise of not only improving glycemic control but also addressing the diverse needs of patients across different populations. Additionally, the integration of oral health into diabetes management underscores the importance of a holistic approach that recognizes the interconnectedness of various health domains.

As we move forward, continued investment in research, education, and healthcare infrastructure will be essential to combat the diabetes epidemic effectively. By fostering collaboration among researchers, healthcare providers, and patients, we can enhance our understanding of diabetes and develop future treatments that not only manage the disease but also improve the overall quality of life for those affected. Ultimately, a concerted effort to address the challenges posed by diabetes will contribute to a healthier society and a reduction in the burden of this chronic condition.

Bibliography

1. Atkinson, M., Eisenbarth, G., & Michels, A. (2014, January 4). Type 1 diabetes. *The Lancet, 383*(9911), 69– 82. https://doi.org/10.1016/s0140-6736(13)60591-7
2. Baxter, M., Hardin-Fanning, F., & Witt, C. (2022, April 20). Expediting dental treatment for patients with diabetes mellitus in a university based dental clinic. *Journal of Dental Education, 86*(10), 1317– 1325. https://doi.org/10.1002/jdd.12944
3. Boles, A., Kandimalla, R., & Reddy, P. (2017). Dynamics of diabetes and obesity: Epidemiological perspective. *Biochimica Et Biophysica Acta (Bba) - Molecular Basis of Disease, 1863*(5), 1026– 1036. https://doi.org/10.1016/j.bbadis.2017.01.016
4. Cefalu, W. and Golden, S. (2015, February 1). Innovative approaches to understanding and addressing health disparities in diabetes care and research. *Diabetes Care, 38*(2), 186– 188. https://doi.org/10.2337/dc14-2748
5. Leslie, R., Palmer, J., Schloot, N., & Lernmark, A. (2015, October 24). Diabetes at the crossroads: Relevance of disease classification to pathophysiology and treatment. *Diabetologia, 59*(1), 13– 20. https://doi.org/10.1007/s00125-015-3789-z
6. Mendenhall, E., McMurry, H., Shivashankar, R., Narayan, K., Tandon, N., & Prabhakaran, D. (2016, June 21). Normalizing diabetes in Delhi: A qualitative study of health and health care. *Anthropology and Medicine, 23*(3), 295– 310. https://doi.org/10.1080/13648470.2016.1184010
7. Oguntimein, O., Butler, J., Desmond, S., Green, K., He, X., & Horowitz, A. (2020, November). Patients' understanding of the relationship between their diabetes and periodontal disease. *The Journal of the American Board of Family Medicine, 33*(6), 1004– 1010. https://doi.org/10.3122/jabfm.2020.06.190454
8. Pang, L., Lian, X., Liu, H., Zhang, Y., Li, Q., Cai, Y., Ma, H., & Yu, X. (2020, August 3). Understanding diabetic neuropathy: Focus on oxidative stress. *Oxidative Medicine and Cellular Longevity, 2020*, 1–13. https://doi.org/10.1155/2020/9524635
9. Siddiqui, N., Sultana, S., Khan, M., Rahman, M., & Islam, M. (2018, December 3). Understanding of diabetes among nurses in a teaching hospital of Bangladesh. *Taj Journal of Teachers Association, 29*(1), 10– 15. https://doi.org/10.3329/taj.v29i1.39086
10. Sinclair, K., Makahi, E., Shea-Solatorio, C., Yoshimura, S., Townsend, C., & Kaholokula, J. (2012, October 20). Outcomes from a diabetes self-management intervention for Native Hawaiians and Pacific people: Partners in care. *Annals of Behavioral Medicine, 45*(1), 24– 32. https://doi.org/10.1007/s12160-012-9422-1
11. Thimm, F. and Szibor, M. (2015, February 22). Mitochondria as a target for future diabetes treatments. *International Journal of Medical Students, 3*(1), 45– 50. https://doi.org/10.5195/ijms.2015.115

Conclusion: Living Well with Diabetes

Managing diabetes is a multifaceted journey that demands a comprehensive approach encompassing education, lifestyle adjustments, and continual support. Throughout this Q&A guide, we have delved into various aspects of diabetes, aiming to provide clarity, practical insights, and empowering information.

Understanding diabetes is foundational to its management. By comprehending the underlying mechanisms, types, and risk factors associated with diabetes, individuals can make informed decisions about their health. This guide has strived to demystify these complexities, offering accessible explanations and expert advice to facilitate a deeper understanding.

Living well with diabetes extends beyond mere management; it involves embracing lifestyle changes that promote optimal health. From adopting balanced nutrition practices to incorporating regular physical activity and monitoring blood glucose levels, individuals with diabetes can proactively enhance their quality of life. This guide has emphasized the importance of these lifestyle modifications and provided practical tips to support sustainable health practices.

Nutrition plays a pivotal role in diabetes management, influencing blood sugar levels and overall well-being. By making informed nutritional choices—such as balancing carbohydrates, choosing nutrient-dense foods, and monitoring portion sizes—individuals can better regulate their blood glucose levels and mitigate potential complications. This guide has offered insights into nutritional strategies tailored to support diabetes management effectively.

Central to navigating the complexities of diabetes is the recognition that support and community play integral roles in fostering resilience and empowerment. Engaging with healthcare providers, joining support groups, and connecting with peers facing similar challenges can provide invaluable encouragement and practical guidance. This guide has underscored the importance of seeking support and utilizing available resources to navigate the ups and downs of living with diabetes.

In closing, managing diabetes is not a solitary endeavor but a collective effort supported by knowledge, proactive lifestyle choices, and a supportive community. Remember, you are not alone in this journey. Embrace the resources and communities available to you, seek ongoing education, and empower yourself to lead a fulfilling life with diabetes. Together, let us continue to strive for wellness, resilience, and empowered living.

Index

C

D

E

F

G

H

I

J

K

N

O

P

Q

R

S

T

U

V

W

Y

www.ingramcontent.com/pod-product-compliance
Lightning Source LLC
LaVergne TN
LVHW080040210926
848481LV00011B/3241
* 9 7 8 1 5 0 1 5 2 7 5 4 8 *